SAMUEL OLDKNOW AND THE ARKWRIGHTS

SAMUEL OLDKNOW

Reproduced from a Portrait in the possession of Col. Wainwright, Marple

Samuel Oldknow and the Arkwrights

The Industrial Revolution at Stockport and Marple

by

GEORGE UNWIN

with chapters by

ARTHUR HULME and GEORGE TAYLOR

and a preface to the second edition by

W. H. CHALONER

Augustus M. Kelley · Publishers

New York · 1968

MANCHESTER UNIVERSITY PRESS
MANCHESTER 13, ENGLAND

Published
in the United States
1968
by AUGUSTUS M. KELLEY · PUBLISHERS
24 East 22nd Street, New York, N.Y. 100010

Printed in Great Britain by Butler & Tanner Ltd., Frome and London

PREFACE TO THE FIRST EDITION

THREE leading contemporary authorities on the early history of the cotton industry and of the factory system—Robert Owen, William Radcliffe, and John Kennedy—agree in attributing considerable importance to the achievements of Samuel Oldknow, who first turned the new spinning inventions to full account by the production of finer cotton fabrics in successful rivalry with the East. In his delightful autobiography, Owen has told us how, soon after he became an apprentice in Mr. McGuffog's shop at Stamford, Oldknow's British Mull Muslins began to displace those of Indian manufacture and were eagerly bought up by the nobility at half-a-guinea a yard. His subsequent account of Oldknow's beginnings as a master spinner and of how "the handsome and imposing mill at Mellor" proved a stumbling block to the ardent young Welshman's earliest ambitions will be found recorded in this book in Owen's inimitable style.

This cotton mill, which passed into the hands of the Arkwright family, was destroyed by fire in 1892, and has since that date been a picturesque and interesting ruin. A detached portion, however, lying by the river-side and within a stone's-throw of the residence built by Oldknow, was not burnt down, and though in a dilapidated condition has been used in parts for stabling or for minor industrial purposes. A body of Austrian prisoners was set to work there during the late war. The distribution of some eighteenth-century weavers' pay-tickets by an adventurous boy scout to casual passers-by led Mr. Arthur Hulme to obtain permission to explore the upper floor of this building, and here on January 1st, 1921, when I joined the search at his invitation, we found a great number of letters, papers, account books, and other business records of every kind and size, covering the whole floor of a large room and partly hidden from sight by several inches of dust and debris. To all appearance the records had lain there for a century, and as the room since the fire had lost its windows they had suffered much from the weather and were in many cases indecipherable. The recovery, cleansing, and classification of the remainder, which filled several sacks, was the work of many week-ends on the spot, and occupied the leisure of the discoverers for months afterwards.

From the first it was clear that by a piece of almost inconceivable good fortune the records of Oldknow's previous business as a muslin manufacturer at Stockport had been deposited at Mellor, so that the documents afforded a unique illustration both of the final phase of the "domestic industry" and of the earliest phase of the factory system. A long and vain attempt to interpret the factory records on the assumption that the fragmentary time books, wage sheets, etc., related only to the Mellor mill ended in the discovery that part of the records were those of a mill simultaneously erected at Stockport, whilst others concerned a bleach and print works previously established by Oldknow at Heaton Mersey; and a further series, which perhaps contributed most to the reconstruction of the whole, related to the very beginnings of Oldknow's enterprise at Anderton, and to the management of the branch there after his removal to Stockport and Mellor.

Inquiries made by Mr. Hulme in Mellor and Marple, where the tradition of Oldknow's achievements is still very much alive, showed that our interest in the factory and its records had naturally been anticipated by some of those whose fathers or grandfathers had been concerned in its working and management, and they kindly placed at our disposal many documents which furnished important links in our investigation. We owe especial gratitude for this fundamental service to Mr. E. Furniss, the agent of the Arkwright estate; Mr. John Taylor, of Marple Ridge; Mr. John Smith, of Marple; and Mr. H. Wheeldon, of Didsbury.

Of the laborious task of bringing into some kind of intelligible order thousands of letters or other separate documents and upwards of fifty more continuous records, of transcribing or summarising the more important, the greater part was undertaken by Mr. Hulme; but valuable help was also given by Mr. G. Taylor, M.A., by Miss F. Collier, M.A., Research Assistant in the Commerce Department of the University, and by Dr. A. Redford, Assistant Lecturer in Economics at Liverpool University, each of whom brought the light of wider research to bear on the documents. Some portions of Mr. Taylor's thesis on the handloom weavers of the Stockport district are included in the present volume and its appendices. For supplementary material drawn from various sources, but chiefly from the advertisement columns of old Manchester newspapers, we are indebted to the generosity of friends who are engaged in investigating other aspects of the social history of Lancashire, and especially to Mr. G. H. Tupling, M.A., of Haslingden, Mr. A. P. Wadsworth, of Manchester, and Rev. W. T. Bushrod, of Chorley.

An attempt was made after a year spent in collecting, arranging, and interpreting this mass of material to give a preliminary account of it in the pages of the *English Historical Review*, under the title of "The Transition to the Factory System." In this article, portions of which with the corrections and modifications necessitated by subsequent research have been embodied in the present book, the important matter of the relations of Oldknow to Sir Richard Arkwright and to his son Richard had to be dealt with conjecturally for want of documentary evidence. This elicited an extremely kind offer of assistance from the late Mr. F. C. Arkwright, the lineal descendant of the great inventor. A few months before his death he invited Mr. Hulme and myself to Willersley and laid open to our inspection the title deeds of the estates at Mellor and Marple, the various mortgages executed by Oldknow, and the deeds relating to the partnership between Oldknow and the second Richard Arkwright which was formed in 1800 and dissolved in 1805. William Radcliffe, in his *Origin of the New System of Manufacture*, says that he acquired his Hillgate premises from the firm of Oldknow and Arkwright. Our records for this period contained no direct evidence of such a partnership, and the statement made in my article that the evidence at our disposal seemed to show that neither of the Arkwrights exercised any internal control over Oldknow's business is to some extent justified by the clause in the partnership deed relieving Arkwright from the responsibility of taking any active part in the concern, whilst a further clause, requiring Oldknow to submit regular accounts, casts a new light on the series of monthly statistical statements which had been one of the most interesting of our discoveries.

Another conjecture made in the article may be regarded as completely disproved by further investigation. The fact that Samuel Oldknow's father was buried at Rivington Chapel, and that his tombstone describes him as "of Nottingham, late of Anderton," led me to the mistaken supposition that the family had migrated from Anderton to Nottingham. Further evidence, kindly communicated to me by the Rev. T. C. Porteus, M.A., Vicar of Coppull, who is investigating the history of the parish of Standish, shows conclusively that the migration was taking place in the opposite direction, and incidentally furnishes another link in the story of Nottingham's early connection with the cotton industry.

PREFACE TO THE FIRST EDITION

In writing the chapter on the Industrial Revolution at Stockport I was very greatly assisted by Messrs. Johnson, of the firm of Henry Marsland Ltd., whose works occupy the oldest and most interesting industrial site in the town, and the title deeds of the Park Mills, placed at my disposal by the firm, have made it possible to trace the continuous history of the transference of the manorial water power from corn-milling to logwood-grinding and silk-throwing, and from silk-throwing to cotton-spinning.

Similar valuable aid was rendered to Mr. Hulme by the Rev. C. G. Mylrea, B.A., who facilitated his access to the parish registers of Marple, and thus enabled him to compile a list of some three hundred and fifty surnames borne by the inhabitants of the district before the building of the factory. This list was essential for the interpretation of the industrial records of over a thousand persons, for which he found materials in wage books and other sources, and which served as a basis for our chapter on the new community at Marple.

That I have been building upon foundations laid by Professor Daniels in his *Early History of the Cotton Industry* will be sufficiently obvious; and I have also to acknowledge much helpful advice from my other friends and colleagues, Mr. T. S. Ashton, M.A., Mr. H. S. Ferguson, M.Com., and Mr. H. M. McKechnie.

For permission to reproduce the portrait of Samuel Oldknow which serves as a frontispiece to the volume we are indebted to Colonel Wainwright, of Brabyns House, Marple; for the drawing of the Methodist Chapel and School to Miss May Malburn; for the drafting of Mr. Hulme's map of the estate at Mellor and Marple to Mr. C. H. Prince, and for the other maps and illustrations to my wife.

A list of the more important records discovered or contributed has been placed at the beginning of the book, and as detailed references to them are impracticable an attempt has been made to indicate on this list the manuscript sources for each chapter of the book. The whole of the records (except those relating to purely local history and such letters as are here published in full) have been deposited by Mr. Hulme, for further investigation, in the Lewis Library of the Department of Commerce at the University, and, along with the records of the historic firms of McConnel, Greg, and Peter Stubs, form a unique body of materials for the history of the Industrial Revolution in Lancashire, Cheshire, and Derbyshire. The publication of researches into each of these sets of records is already in progress, and it is to be hoped that other old firms who possess similar title-deeds to a place in history will deposit them in the archives of the University.

George Unwin.

Christmas 1923

PREFACE TO THE SECOND EDITION

SINCE Unwin published *Samuel Oldknow and the Arkwrights* in 1924 our knowledge of the early factory system in cotton and of the problems to which it gave rise has expanded considerably,[1] but little new work has appeared on Oldknow himself and the business papers which he left behind. The publication in 1949 of Stanley Pigott's *Hollins: a Study of Industry* by William Hollins & Co., Ltd., of Viyella House, Nottingham, revealed much interesting information about the Pleasley mill in which Thomas Oldknow (1744–1817), Samuel's uncle, was an original partner in 1784.[2] W. H. Chaloner's "Robert Owen, Peter Drinkwater and the Early Factory System in Manchester" (*Bulletin of the John Rylands Library*, vol. 37, no. 1, Sept. 1954, pp. 78–102) shows that for this period Owen's autobiography, *The Life of Robert Owen written by himself*, vol. I, 1857 (on which Unwin leaned very heavily), is chronologically inexact.[3] Unwin did not know that the famous mechanical

[1] See, for example, N. J. Smelser, *Social Change in the Industrial Revolution: an application of theory to the Lancashire cotton industry, 1770–1840* (1959); Frances Collier, *The Family Economy of the Working Classes in the Cotton Industry, 1784–1833* (1965); S. D. Chapman, *The Early Factory Masters* (1967); and the bibliographical introduction to the second edition, 1966, of E. Baines, *History of the Cotton Manufacture in Great Britain*, originally published in 1835.

[2] There are several errors in the family tree of the Oldknows printed at the end of Mr. Pigott's volume: Samuel Oldknow I (1733–59), not our Samuel (1756–1828), married Margery Foster; Samuel Oldknow II (1756–1828) never married, as far as is known; and Thomas Oldknow III, draper, of Nottingham, Samuel II's cousin, is entirely omitted. He committed suicide in 1812 after murdering his two young children (*Manchester Mercury*, May 12th, 1812, p. 4, col. 6).

[3] The necessary revisions have been made to pages 152 and 155 of Unwin's text. These are the only changes in the volume.

engineer, Peter Ewart (1767–1842), was in partnership with Oldknow for about twelve months in 1792–3, and the following extract from W. C. Henry, "A Biographical Notice of the Late Peter Ewart, Esq." in the *Memoirs of the Literary and Philosophical Society of Manchester*,[1] is of extraordinary interest as throwing light on the financial crisis of that period and its effect on Oldknow's fortunes:

"In the year 1792, Mr. Ewart was induced by the liberal proposals of Mr. Oldknow to relinquish his profession of Civil Engineer, and to enter into partnership with that gentleman, in the Cotton Manufacture. At this period, he became known to the late Dr. Currie, of Liverpool, the illustrious biographer of Burns; and in the following interesting letter to Mr. Wilberforce from Dr. Currie, is placed on record the deliberate judgment, which that sagacious and discriminating observer had formed of Mr. Ewart's talents and character.

'Dr. Currie to W. Wilberforce, Esq.

LIVERPOOL, *23rd April*, 1793.

'Dear Sir

'If in the long letter, which I wrote you two days ago, there appears a good deal of unguarded warmth, the following circumstance will explain, though perhaps not justify it.

'I was sitting in my study, on the evening of Saturday, reflecting on public affairs, when a young man called to drink coffee with me, a manufacturer of Stockport, near Manchester. After giving a picture of the general distress

[1] 2nd series, vol. VII, 1846, pp. 120–5. For further details of Ewart, see H. W. Dickinson and Rhys Jenkins, *James Watt and the Steam Engine*, 1927, pp. 288–9 and *passim*; F. Greenaway, *John Dalton and the Atom*, 1966, *passim*.

there, he informed me of his own situation in particular, and of the business which brought him to Liverpool.

'He said that the house, of which he is a partner, employed about 1,500 hands, all of whom are now idle, or, as the phrase is, off work. That previous to their being discharged, he and his partner had struggled on from one week to another, in hopes that the times would mend, and a demand, more or less, come for their goods. That, in this hope, they had gone on for the last three weeks, and not having a sufficient quantity of money to pay the people their full weekly wages, they had prevailed on them to accept about a third of the sum, as this, with economy, might suffice for subsistence. In procuring the money for this purpose, he told me, they had been reduced to extraordinary difficulties. Formerly they sold their goods in large quantity, but now they determined to supply the retailers themselves with a single piece, or even less ; and, provided they paid them in specie, at almost any price.

'Accordingly, having goods in their warehouses that suited the home market, they fitted up a light cart and sent a young man with it, full of goods, to supply the retailers in every part of the country, and to bring home the specie every Saturday, whatever might be the loss. The expedient succeeded for about three weeks, but had now failed, and he was come to Liverpool to try if by any possible means he could raise a few hundred guineas, to get over another week and keep his people alive. He told me that he and his partner had been constantly amongst them, and by entering into all their distresses, had prevailed on them to be extremely patient and reasonable. At their last meeting, they had agreed to wait this young gentleman's return from Liverpool, and what money he was able to raise, they had consented should be laid out in oatmeal, which being boiled up with water, potatoes, and some of the coarser peices of beef, should be shared out in fair proportions among them ; and thus in the cheapest manner provide for their subsistence. As the house had many thousands owing them in Liverpool, though he knew there was no hope of any considerable debts being paid, he had no fear of not being able to procure the sum immediately wanted. He had been using every effort for two days, and had actually threatened to arrest two of our principal merchants on the exchange, but he had not been able to raise a single guinea. How he was to face the poor people he knew not, each of whom had four to six weeks' wages due. But he could appeal to Heaven for the anxious exertions, which he had made, to relieve distresses which he could neither foresee nor prevent. As I looked at this young man, I perceived that his countenance seemed actually withered with care and sorrow. He is not a common character ; he was the apprentice of Messrs. Boulton and Watt, and had an extraordinary degree of the most useful knowledge of every kind. He is modest, virtuous, and prudent ; of astonishing application, and, in a word, one of the first young men I ever knew. These qualities recommended him to the notice of the manufacturers, among whom he exercised his profession of a mechanic and engineer. He had offers of partnership from the first houses there, and was actually taken into the house of Mr. Oldknow, of Stockport, about a year ago, at that time perhaps the first establishment in Lancashire. Mr. Oldknow you must have heard of, as the original fabricator of muslin in this country, and a man of first-rate character. He has laid out a property of £50,000 on building and machinery alone. His partner, (the young gentleman I speak of,) is named Ewart, the younger brother of Mr. Ewart, the late Envoy at Berlin. It is men such as these that are reduced to such extremities.'

" In consequence probably of this fearful depression of commerce, surpassing evidently in extent, and severity any subsequent crisis, Mr. Ewart's connexion with Mr. Oldknow was dissolved after the lapse of a year."

Criticism of the views of Unwin and his collaborators on Oldknow's community building at Mellor and Marple has come from S. Pollard.[1] On the question of community building by the early factory masters, Professor Pollard accepts that " each entrepreneur, when faced with the pressure of rapid decision making, was bound to draw on the deeper resources of his character to solve each immediate crisis,"[2] but considers that Oldknow was " merely fortunate in his biographers," and that their claim that ambition drove him " to control and direct the life of the community " is without foundation. On the contrary, writes Pollard, "all the records . . . point to a simple decision to establish a cotton mill on a village site, with only a belated recognition of the community needs involved."[3]

While it must be admitted that no statement by Oldknow of an intention to found a colony has ever been discovered, it is quite clear from the evidence brought forward in chapter IX of *Samuel Oldknow and the*

[1] " The Factory Village in the Industrial Revolution," *English Historical Review*, July 1964, pp. 529–30.

[2] *Art. cit.*, p. 530.

[3] *Art. cit.*, p. 530. Pollard gratuitously adds : " Nor is there any evidence that he was anything but a mediocre businessman, who never understood the reason for his windfall profits in the seventeen-eighties and his bankruptcy in the seventeen-nineties " (*art. cit.* p. 530, fn. 1). But unless business success is to be one of the main criteria governing the choice of subjects in the writing of social and business history, it is difficult to see the relevance of this statement.

Arkwrights that the formation of the rural-industrial estate at Marple and Mellor had complicated origins extending over the period from 1787 to 1791, with more to it than a " simple " entrepreneurial decision. And as Professor Ashworth has pointed out: " Financially, it [Oldknow's attempt at community building] was not a very successful experiment, mainly because those enterprises which arose from Oldknow's desire to build a community, and which had no necessary connexion with cotton-spinning, absorbed an amount of capital on which he could not afford to continue paying interest."[1]

The Oldknow papers used by Unwin and his fellow researchers are now somewhat scattered. The largest deposit is in the John Rylands Library, Manchester.[2] A second deposit is in the E. R. A. Seligman Collection in Columbia University Library, New York City. Some of this material is available on microfilm in the Manchester Central Reference Library. Other documents relating to Oldknow have been carefully and enthusiastically collected and copied by the Marple Antiquarian Society and are available on microfilm in the Manchester Central Library.

Finally, certain other private documents relating to Samuel Oldknow are still in the possession of his legal heirs.

W. H. Chaloner

The University of Manchester
June 1967

[1] W. Ashworth, *The Genesis of Modern British Town Planning* (1954), p. 120.

[2] See Moses Tyson, " Handlist of Additions to the Collection of English Manuscripts in the John Rylands Library, 1928–1935," *Bulletin of the John Rylands Library*, vol. 19, 1935, pp. 49–60 [Ref. no. R73730, 751–840].

CONTENTS

CONTENTS—*Continued*

LIST OF ILLUSTRATIONS

LIST OF THE MORE IMPORTANT RECORDS FOUND AT MELLOR

MUSLIN MANUFACTURE AT ANDERTON AND STOCKPORT

Chapter I.

Ledger, 1782–86, containing accts. of creditors and servants, not of customers.
Account Book, 1782–4, containing Day Book of Sales, 1782–4; Cash Book, 1783; Balance Sheet, Stocktaking, List of Weavers and Spinners for 1783; Petty Expenses of S. O., 1783–4.
List of Yarn Deliveries in 1784, on sheet of brown paper.
Accounts of Carriers, Bleachers, etc., for 1782–4.

Chapter III.

Weavers' Ledger, complete with exception of occasional pages for most of 1784–5, fragments of other ledgers for 1786 and 1790.
Weavers' Taking-in Book from Nov. 1786 to May 1787.
Winders' Ledger, 1788
Day Books of Sales for 1786–7, 1788, 1789–90.

Chapters IV. and V.

Letters of S. Salte, R. Arkwright, S. and T. Oldknow, 1786–88.

Chapter. V.

Spinners' Ledger, 1787–8 : Accounts of about 50 small spinners in Stockport District.

Chapter VII.

Ten Warping Books, 1787-94 : A record of the daily output of 13 mills.
Finishing or Cutting Frames Book : Record of work on 30 frames, March to July 1787. Also Time Book of about 50 girls.
Fragments of Daily Accts. of Sizing and Finishing Output for 1788.
Costing Books : One fairly complete, and fragments of another, undated.
List of Piece Goods wanting places in Sales, 1788.
Stocktaking Records, 1789–93.
Reeds and Healds in Weavers' Hands without Work, 1789. Stock of Weft.
Stock of Materials in Weavers' Hands, Aug. 24th, 1790.
" Stock of Weft and Twist, 1790." Stock of Weft, Twist, and Warp, no date
" Stock taken Sept. 1st, 1791 " : Gives an account of the materials on each of the machines in the newly-built Hillgate factory. Also Stock of Materials at the Carrs Shop. (Chap. VIII.)
" Stock of Reeds and Gears, Aug. 1st, 1792 '
" Stock taken July 1st, 1793 " : Piece goods, twist, and weft.

LIST OF THE MORE IMPORTANT RECORDS FOUND AT MELLOR—*Continued*

Chapter VII.—*Continued*

List of Weavers, with Materials in hands of each, 1793 : Another list classified according to their products.
Prices of Weaving, March to June 1791.
Weavers' Pay Tickets, 1789–94 : A great many daily bundles ; two complete months in 1794.
Lists of Goods delivered from Croft, 1791–3 : Separate sheets, each with a day's record.
Anderton Fortnightly Accounts of Payments to Spinners and Weavers.

SPINNING AT STOCKPORT

Chapter VIII.

Carrs Spinning Accts. Five Wage Lists, Oct. 1789—Feb. 1790.
" Sent to Mr. Oldknow," a daily account of output, May 1790—Oct. 1791.
A Book of Orders for low counts of yarn.
A Cotton Deliveries Book.
Time Book for Hillgate Factory for most of 1792 (does not include piece-work spinners).
Spinners' Weekly Accounts for piece-workers, Mar.—Nov. 1793.
" Manufacture to Spinning, Dr. Acct.," Dec. 1791 to Dec. 1793.
Stockport Cotton Pickers' Wage Book, 1790–1.

Chapter X.

Letters of Richard Arkwright junior, S. Salte, H. Norris, 1791–4.

SPINNING AT MELLOR

Chapters XI. and XII.

Time Books of Workers in Mellor Factory, 1793–4 and 1794–5.
Two Pickers' Wage Books, 1792 and 1795.
Shop Notes for 1793–4 (detailed account in Chap. XII.).
Book of " Notes Delivered Out," with dates of return, Dec. 1793 to April 1794.

Chapter XIII.

Statistical Statement of Output for January 1797, with Estimate of Costs.
Daily Statistics of Output from May to Sept. 1797.
Stock of Cotton in Mill, Sept. 1st, 1798.
Fragment of Ledger with leading accounts for 1801–4.
Two detailed Output Books for Years 1806–1812.
Twenty-one Monthly Sheets of Mill and Estates Statistics between Feb. 1804 and Feb. 1807, numbered 61 to 100.
Four Sheets and Two Books of Fortnightly Wage Accounts : the sheets relate to 1803–4–5 ; the books to June 1805—April 1806, and Oct. 1810—July 1811.
Nine Monthly Sheets of Output and Cost Statistics, for Dec. 1811 and Jan. to

LIST OF THE MORE IMPORTANT RECORDS FOUND AT MELLOR—*Continued*

Chapter XIII.—*Continued*

May, Oct. to Dec. 1812, on same pattern as that for Jan. 1797.
Nine Monthly Sheets of Wage Statistics—two for 1811, seven for 1812.
Three Pocket Memorandum Books with notes of debts, orders, and contracts, 1803, 1805, 1807

MISCELLANEOUS MELLOR AND MARPLE RECORDS

Chapter IX.

Detached Accounts of Farming and General Labour, 1787–90.

Chapter XI.

Wages Paid at Mellor: 18 weekly and fortnightly lists, 1791–3.
Three Flesh Books, 1792–3; Three Coal Books, 1792–5; Three Gardening Books, 1804–09–10. Contain accounts of supplies to workpeople and others.

Chapters XIV., XV., and XVI.

Book containing Directions for Outdoor Labour, Farming, Stone-getting Building, etc., 1797–8.
Book of all Wages for Outdoor Labour for 1799–1800.
Wage-tickets for Outdoor Labour, in fortnightly packets, spread over the years 1801–12.
Canal Permits, 1797–1805.
Canal Accounts.
Canal Toll Accts.
Wage Book of Outdoor Workers, 1809.

Chapter I.

MUSLIN-MAKING AT ANDERTON 1782–4

IT has generally been claimed by those who have won wealth or distinction in commerce or industry that they were of yeoman origin, and as the term yeoman has been employed loosely for many centuries to cover all enterprizing holders of land who were not gentlemen, serfs, or labourers, the claim might be made out in many cases if the facts were known. But as most Englishmen of whatever origin who have made money in trade have become land owners, and as it was quite common down to the 18th century to carry on farming in conjunction with a town business, the reference to yeoman origins as a basis for large social or economic generalization is of dubious value. The close connection between the two fields of enterprize is however an extremely important aspect of social and economic history which will receive ample illustration in the course of our narrative.

Thomas Oldknow, the grandfather of Samuel, died in 1787 at an advanced age at Calverton, a village some six miles to the north of Nottingham, and it is not improbable that he was born there and held some land. But it is fairly certain that in middle life he carried on business in Nottingham and very probable that the draper's shop kept by his son Thomas in 1780 had been established by him sometime before 1750.

About that time another son, Samuel, the father of our hero, was probably sent to Anderton in Lancashire to learn the business of a manufacturer, as we find him in 1754 at the age of 20 described as a tradesman of that place, in the register of his marriage with Margaret, daughter of Thomas Foster of Anderton. At Roscow Low, one of the leasehold estates inherited by his wife from her father, who died soon after, Samuel Oldknow carried on his business and erected a new building for the purpose, which still bears his initials and the date 1759.

In that year he died at the early age of 25 and was buried at Rivington Chapel near Anderton, where there is a tombstone bearing the inscription "Samuel Oldknow of Nottingham, late of Anderton." The widow remained on her estate at Roscow Low with her three children: Samuel (born in 1756), Thomas, and Elizabeth, and later on married John Clayton, a farmer. The records give us many glimpses of the Clayton family. We hear of three children: Margery, Sam, and John. The last-named joined Samuel Oldknow at Mellor and succeeded to the management of the mill after his death in 1828.

In course of time young Samuel was apprenticed to his uncle Thomas at Nottingham. In 1781, when in his twenty-fifth year, he was taken into partnership with him, and shortly afterwards it was proposed that, whilst retaining his connection with the Nottingham business, he should return to Anderton and set up as a manufacturer of cotton goods and fustians.

The Nottingham shop would furnish the beginnings of a market, but there was to be a saleroom in Manchester, and Samuel might, as a further string to his bow, undertake an agency for Nottingham hosiery. There can, however, be little doubt that the production of muslins, which within eighteen months had become the essential feature of the enterprize, was from the first under consideration. Professor Daniels in his *Early English Cotton Industry* has shown how favourable in 1781 the conditions were to such an undertaking. The disallowance of Arkwright's patent for carding as a result of the trial of that year, and the almost simultaneous publication of Crompton's invention, gave an immense stimulus to the manufacture of the finer cotton fabrics.

MUSLIN-MAKING AT ANDERTON, 1782–4

It took me [says the inventor of the Mule] from 1778 to 1779 to finish it. From 1779 to the beginning of 1780 I spun upon it for my own use both warp and weft. In the beginning of the year 1780 I began to spin only and left off weaving . . . I had not used it constantly more than Six Months before I was beset on every side by people of various descriptions from the distance of 60 miles and upwards as well as my neighbours . . . whose curiosity was excited by the superior quality and fineness of the yarn I spun hitherto unknown, and which at that time the trade was much in want of. In the end of 1780 it was made public.

The Mule was first known as the Muslin Wheel, and the inventor himself later described it as "that piece of mechanism that has produced and increased one of the first manufactories in Europe, viz., the fine Muslin and Cambric."[1]

To succeed in the production of muslins had long been the highest ambition of British manufacturers. An attempt had been made at Paisley in the days before the Union, but it had failed to meet the test of Indian competition. There is a tradition that the first English muslins were woven at Anderton in 1764, by Joseph Shaw, and this suggests the possibility that the business set up by Samuel Oldknow the elder in 1754 had been connected with an experiment of this kind. If so, the enterprize of his son in 1782 may be regarded as suggested to some extent by filial piety. But the emergence in the Bolton district of a supply of fine yarn, for the want of which previous attempts had failed to achieve permanent success, must have prompted many ingenious persons to renewed experiments, and it would be as difficult to disprove as to prove the claim made in the *Manchester Courier* for April 18th, 1829, and renewed in the *Bolton Journal and Guardian* for January 13th, 1922, that the

[1] Daniels : *Early English Cotton Industry*, pp. 168-9.

muslin manufacture was permanently established by Thomas Ainsworth at Bolton in 1780. All that can be claimed for Samuel Oldknow is that he first achieved eminence in the industry. In the last half of the 18th century a draper's shop often provided a stimulus to new industrial enterprize. Robert Owen and David Dale shared this starting point with Samuel Oldknow, and many of the founders of the great warehouses of Manchester had received the same training. Owen attributed the success of his former employer McGuffog, who had commenced as a travelling packman, to his skill as a buyer rather than as a salesman. It was the draper's business to find out where the best goods were being made, to stimulate their production, and to secure the handling of them. In this way he might become either a manufacturer like Samuel Oldknow or develop a great warehousing business like the Potter Brothers of Manchester, who set out about the same time from a similar draper's shop at Tadcaster.[1]

The fortunate survival of a number of letters rescued one by one from the factory debris at Mellor, and of two of Oldknow's earliest account-books with only a few pages missing, enables us to follow in some detail his operations as a manufacturer from his settlement at Anderton about Midsummer 1782. The letter bearing the earliest date supplies the best introduction to what follows and sheds that kind of light on the subject which the historian is always seeking and seldom obtains. It is from a business friend of the Oldknows in Manchester whose advice has been asked concerning the proposed enterprize. But if the ambition of making muslins had been already conceived, it was thought wiser to conceal it.

[1] E. Meinertzhagen: *From Ploughshare to Parliament.*

Samuel Hardman to S. Oldknow:

"MANCHESTER, *December 22nd*, 1781.

"I duly received your favor of the 2d Currnt and observe you have come to a final resolution to embark in the Cotton Manufactory—in which I sincerely wish you success. As you say, you certainly will have many difficulties to encounter, and I think your first Plan shou'd be to engage an *honest Man* for a year, or as long as you think proper, that understands the manufactory well in all its different branches, and as soon as you can by application and attention to make yourself Master of the Business. You mention you intend to divide your time between Anderton and Manchester. I think one day in a Week wou'd be enough to Stay in Manchester, at present, and you might save some expence by being at Anderton the remainder of the week, besides being in the way to attend immediately to making of Gray Goods. You mention you cou'd Wish a Warehouse and Lodging together near the Center of Business, but I think you are not apprized that such an one as you describe wou'd be a very high Rent—and tolerable Warehouses are very difficult to be got—but shou'd I see anything I think wou'd suit, will advise you.

"I think your intention of selling Hosiery Goods wou'd not answer, few of them wou'd be sold here, and to a Person that bought your own Manufactory wou'd have an appearance as if you dealt in many things foreign to the Cotton Trade— . . . The Fustian Trade is often varying tho' the first Principles are much the same."

The first six months were obviously a period of experiment. The weavers whom he was recruiting for his muslin manufacture had at first to be employed on the articles they were accustomed to make. These were the fancy cotton goods of which James Ogden in his famous description of Manchester in 1783 has attempted an enumeration—Velverets and Velveteens, King's Cords and Queen's Cords, Herring-bones and Buff Jennets, Dyed Pillows and Waistcoat Jeans, etc. Not until November does the despatch to several customers of three pieces of Ballasore handkerchief suggest that Oldknow's career as a muslin manufacturer is beginning. During the three months from September 24th (when the record of sales begins) to Christmas about £400 worth of piece goods were sent out. Half of these were disposed of by the Nottingham shop and by a firm of linen drapers in Mansfield, and one of the remaining quarters was taken by Mr. Samuel Mather, who is probably the silk and fustian manufacturer of King Street mentioned in the Manchester Directory of 1788. As Oldknow spent far more on cotton than on yarn at this time, he must have found work for a number of small spinners in the neighbourhood of Anderton. He likewise opened an account with Messrs. Peel and Yates for printing and with a fustian calenderer for the finishing of his goods.

In the beginning of 1783 a new epoch opens in the affairs of Samuel Oldknow. The partnership of uncle and nephew ceases, though the uncle continues to have important financial relations with the nephew and to offer friendly advice. The chief market for the goods produced is no longer found in the Nottingham connections or even in Manchester. From this time onwards two London firms, Messrs. S. & W. Salte and Messrs. Parker, Topham & Sowden, take about two-thirds of the rapidly increasing output of Samuel Oldknow's manufacture. But—most important of all—it is from the spring of 1783 that Oldknow becomes primarily a maker of muslins. Within three years he was recognised as the first in the kingdom.

Though the detailed record of his sales for this year is wanting, a ledger containing the accounts of his creditors, an account-book containing a cash account of receipts and payments, a full inventory of his stock-in-trade in August 1783, a list of the spinners and weavers in his employ, and an account of his personal expenses furnish full particulars of the expansion of his business, and enable us to form a sufficiently clear idea of its commercial and industrial organization. But the central clue to the reconstruction that is taking place is to be found in Oldknow's correspondence with the two London firms above referred to.

Oldknow, who had already improvised a warehouse in Anderton in a building adjacent to the house of his stepfather, had entered, in January 1783, into occupation of a salesroom in Manchester on the premises of Mr. Cririe, a merchant in St. Ann's Square, at a rental of £13 a year. Here he accumulated a stock of his goods for show, and rode up weekly at the recorded expense of four or five shillings to push sales and to buy cotton weft and twist. But no sooner were his weavers adequately trained, and a steady flow of muslin products begun, than he discovered that London, and not Manchester, was the most effective market for his wares. The travelling partners of the two firms of Parker, Topham & Co. and S. & W. Salte, who were eagerly looking out for the latest novelties in Manchester goods for the spring trade, had prospected a gold-mine in Oldknow's muslins, and each of them was offering to take more than he could produce. Oldknow was in a dilemma. His artistic tastes and his impulsive and speculative temperament urged him to throw himself unreservedly into the manufacture of muslins, but apart from a possible lack of fine yarn and skilled labour there were two serious obstacles to the expansion of the business. His output was limited by the smallness of his capital and credit, and the muslin manufacture was liable, not only to all the fluctuations of a seasonal trade, but to a sudden and severe burst of competition whenever a large cargo of Indian muslins came into port. Whilst, however, he needed considerable advances of capital and the guarantee of a steady market for his goods, he did not wish to purchase these advantages by a complete sacrifice of his independence. A draft of a letter to S. & W. Salte on April 22nd shows him cautiously feeling his way in this negotiation. Each of the two firms was willing to make advances of capital, but each bargained for an exclusive agency. Oldknow compromised the matter by dividing the London agency between them, and they continued to take two-thirds of his entire output for a number of years.

Draft of letter in S.O.'s handwriting :

" Mr. Mower,

" Sir,

" Pardon me Sir, I take this liberty of requesting of you some friendly information, but previous [to my saying of what nature it is] it may be proper to say who I am. I am the same Samuel Oldknow that was in partnership with Thos Oldknow of Nottingham in 1781 & 1782 and now I am a resident of Lancashire in the Cotton Manufacture [just beginning to sell the produce of my manufacture]. Muslins is an article I have made. I have recd. considerable orders from your late partners Parker & Topham and have sent 'em to the amount of £150. It is a great satisfaction in business to know the ground one acts upon and I shall esteem it a particular favour you will inform me whether Parker and Topham are safe people to do business with. Whatever you think proper to communicate to me on this head you may depend on silence with respect to your name."

T.O. junior to his nephew S.O., probably in March or April 1783 :

. . . I have been in London again . . . to you last on Bro Johns affairs. Sam Salte desires you will send him Samples of what you have coming forward as soon as you can. I am off opinion he will take all you can manufacture, if you can make them like what you sent mee now. With regard to what you hint about the House in Watling Street I cannot say anything about it. I only know that whenever I have calld at the warehouse he hath always been on the spot & I always think that hath a good look.

"I shall send you this week by way of Derby to be forwarded by coach to be at Manchester on Saturday Cash £100.

"I purpose being in Lancashire some time this month. I am not in want of many goods and I think it will be better in the print line rather later . . .

"With regard to the Bond to Mr. Abraham Crompton, you had better get a bond filled up to his own mind leaving out the dates till it is executed."

S.O. to S. & W. Salte (?) (draft):

"ANDERTON, *April* 22, 1783.

"GENTLEMEN,

"By this days coach I have forwarded you a small assortment of Muslins (invoice inclosed) agreeable to the directions you was so obliging to give to my Uncle on my account. I am sorry it has not been in my power to send them sooner. I have not sold three times more than the quantity I have now sent you and they are principally gone to a house in London who visited me hear previous to the receipt of your order. They want to take all my manufacture. I have just recd. a letter from them which seems to be written in anger because I do not send them fast enough. I am equally as much attached to the service of your house as theirs, but upon my present plan of my Muslin Manufacture I cannot serve anyone with many. (((I will just speak to the difft sorts of goods I make and to the present situation of my Muslin Manufacture with a view to have your opinion on the british Muslins and that I may judge whether it will be prudent to go wholly into that manufacture, for [to] do it capitally it will very much alter my mode of proceeding.)))

" To do anything capital in the Muslin business I must very much alter my mode of proceeding. As yet I make other articles in the fustian line. Now if the Muslin trade will answer at all it is of that magnitude as will engage all my attention as (((as five times more money than I))) a business of itself and to go into that line entirely—to serve a London connexion—I should have no occasion for a warehouse at Man'r or to give a day or two attendance every week there (((upon the sale of goods and to give up these one ought to))). If I could be certain of the Muslin trade continuing with us hear I shd not require a moment to determine what to do. The prospect is at present very propitious (((but at a time when East India Muslins are exceeding scarce and in all probability will not long continue so—it may not be))) but how will it be when East India Muslins are more plentiful. I do think there are some sorts of very broad muslins we have the lead in and in all others we are daily improving—fine spinning is what we are most shot of & even that we are on the road to procure. The finishing part is what we have long been striving to master and in this I wish you to say what you think of mine. I do them myself and the people I employ in that department are under an obligation not to disclose the secret. It is in the ydwds [yard-wides] you are to judge by for in them we have made an improvement in the selvages. I shall esteem it kind you will give me your opinion on the whole as soon as convenient for what you say on the subject will be an important argument with me on steping further into the business. I take very great delight in the manufacture and shall always be anxious to excel. I am not making any more 5/4 Stripes, 6 & 7/4 is what I mean to make in future—the narrow cord is not fine enough to answer—it is very expensive and I intend to drop making unless you think such an article will take, Waiting your answer, I am—

((( ))) = crossed out.

S.O. to S. & W. Salte (draft):

" *May* 1783.

" Your esteemed favor on the 26 of last month I recd. in due course in consequence of which I have sent as by Bill annexed. I have gone this afternoon to Manchester and will forward them to morrow as at foot—6/4 & 9/4 Stripes and Checks I have not any ready, indeed I have very few made—but I am setting more Looms to work as fast as possible in order to supply you. I have added 4 demi small checked Cords more than you ordered—if you think that article will not sell—will thank you to say so in your next and I will certainly drop the making of them—You may depend on my always doing the best in my power for you—at present materials are too high to make Muslins very low. I buy no warps at any cheaper rate than of Messrs. Arkwrights & Co. & their Twist is too dear to make them at any lower prices. I see a way to procure materials that will at least render each ps at 7/– less—but cannot accomplish my design without a little time—I thank you for the priviledge you allow me in Drawing upon you, and when I have occasion to make use of it will regularly advise you—

" I wish for a personall conversation with you on the subject of Trade, Muslins in particular, and if my business will allow me a fortnights absence my intentions is fixd for a journey to London in the Month of September."

S.O. to Parker & Topham :

"ANDERTON, *May* 17, 1783.

"DEAR SIRS,

"Your favor of the 12th inst., I have this day received. You say you have made some remarks on some of my Handfs being charged rather higher—I do assure you I have sent you none but what I believe would have sold for 7½ pr ct in the Manchester Market, nay I am certain of it. . . . you expect I am sending the whole of my manufacture to your house and on that account you have declined an offer. I shd recommend you to accept of any offer that is likely to be of advantage to you—I do send you a considerable share of my manufacture—but I cannot engage the whole to your house—this I will promise you that no one shall be charged lower than you are. I thank you for the liberty you allow me in drawing upon you. I have a few of every kind coming forward but I cannot get supplyed with fine materials to go on with—or should be able to send you more—this I trust will not be the case long."

S. & W. Salte to S.O. :

"LONDON, *May* 24, 1783.

"We rec'd the muslins and neckcloths etc & also your letter—to make short of the business we wish you could spare time to come to Town—we can do more by conversation than letter & settle all our future plans & be assured we mean to make it a Mutual Interest. We have a Muslin Sale next month—we can, if you come to London, shew you Patrns and give you directions which a letter cannot convey properly. With respect to any asistance in our power, you may comand it—we wish to encourage merit wherever we find it—but we profess a personal attachment & Friendship for you. Do send the 6/4 & 7/4 chks & stripes & as many neckcloths as you can. We rather wish to take the whole you make, as another person interfering may be of no advantage to either. Give us an answer soon & and let us know when we may expect to see you."

T. Oldknow Junior to S.O.:

"NOTTINGHAM, *May* 14, 1783.

"DR. NEPHEW,

"I have been with Mr. Robinson and he says he cannot send any twist till next carrier when he will send you a part off the order by Spencer Car. paid to Mancr. He says building &c hath swallow'd so much money that at present he sells for ready cash. I told him he might have the cash when the goods was sent off allowing the usual disc't & that if the article suted, you would be glad to be supply'd weekly which he could do."

Having learnt something of the external relations of the new enterprize, we may now turn to consider its internal economy. Ample data for this purpose are provided by the recorded stocktaking of August 1783. But first as regards the warehouse at Anderton, in which the business was centred in 1783, there is an interesting letter, undated but clearly belonging to a later period, asking the landlord for a lease on the ground of improvements effected by Oldknow, probably in 1784. A plan is attached to the letter showing that the original structure adjoining Mr. Clayton's house was 31 feet 2 inches long by 19 feet 6 inches wide, and that the addition made by Oldknow had practically doubled the accommodation at a cost of £90. That an eminent manufacturer who claims to have established a new industry should regard £90 as a considerable outlay shows how small a part fixed capital in buildings or machinery as yet played in industrial enterprize.

S.O. to Mr. Gill (draft) no date :

" I presume you are not unacquainted with the Building I have built and the alteration I have made at Anderton for the purpose of a Warehouse—the money expended is £90—you will say too much for me to lay out on any Gentleman's Estate with out some consideration—the matter stands just as follows, before I began to build I applyd to Mr. Hawkshead for a lease of the premises either for the life of Marjery Clayton (my Mother's Daughter) or for some fixed number of years—the same he made known to Mr. Blundel and said to me he had obtained of him a promise to grant me a lease of the Premises for 21 years—me paying a trifling acknowledgement to Mr. Blundel the Lord of the soil—this I accepted and went on with the Building—Many times I applyed to Mr. Hawkshead to get the lease made and executed and furnished him with the ground plan of the building (which I now give you on the other side) I might not be so urgent to get this work done as many would have been on account of living at so great a distance, being so much engaged in Business and never doubting I should have it very soon.

" This is the present state of this business *nothing* done, therefore I have to beg of you that you will speak to Mr. Blundel on behalf of my interest, to obtain leave to get lease made for 21 years, paying 5/– acknowledgement to bear date from January 1st 1784. I have not a personal acquaintance with Mr. Blundel or I would have wrote to him upon this business myself—I do not think he will hesitate to grant me a lease agreeable to the first intent & meaning of my laying out so much money."

(((If I understand Mr. Blundel right I am sure he will not. If he does you may safely say he should not to me at least one who has been a principal assistant in the establishing a manufacture in this kingdom, whereby the land lords property will so much advantage—tell him I hope I have made the whole of his Estate of as much more value as the Roscow Low Estate is worth.)))

((( ))) = crossed out.

Although there was not much room in Oldknow's warehouse before it was enlarged, it is conceivable that it sufficed for his stock-in-trade at that time. The whole of his "Fixtures and Utensils," which comprized office furniture as well as machinery, was valued at £57 17*s.* 11*d.* This modest equipment covered all the processes of preparing the warp; giving out the cotton, warp and weft; taking in, examining, finishing, and storing the cloth.

The warehouse, even before it was enlarged, had apparently separate rooms for giving out and taking in. In the former were four bags of cotton of different kinds, the total quantity being only 324lbs. The cotton of Berbice (77lbs. at 2/–) had hitherto been thought the finest, but that of Brazil (130lbs. at 1/9) was now thought to be equally good material for fine fabrics. There were also 97lbs. of St. Domingo at 1/7½ and 20lbs. of Smyrna for coarse fabrics at 1/4. Besides this there were 125lbs. in the hands of 23 small spinners employed by Oldknow, and a further quantity given out to a score of weavers who saw to the spinning of their own weft, making a total value of about £50. Carefully sorted out in the cupboards and drawers of this room, there lay about a couple of hundredweight of weft and twist of all counts from twenties to nineties, wound or unwound, and valued at £180. The nine winders had 80lbs. more, but most of the yarn was in the hands of the weavers. Of

[1] *Cf.* Ogden's *Description of Manchester*, p. 88, and Daniels' *Early Hist.*, p. 38.

the sixty-nine weavers employed by Oldknow one was debited with only 2½ lbs. of cotton and three with a warp apiece. Of the remainder, two-thirds had received warp and weft and one-third "warp and wool." In a score of cases the weavers are said to have received with their weft or "wool" only a third, a half, or two-thirds of the warp, the implication being that they were to provide the rest themselves. The total value of the materials thus given out to the weavers was £261 17*s.* 11*d.* But by far the greatest part of Oldknow's assets was to be found in his stock of manufactured goods valued at £812, of which four-fifths were muslins mainly in the Anderton warehouse, and one-fifth other cotton goods mainly in the Manchester salesroom; and in the book-debts owing to him, amounting to £1,407. Against total assets in these various forms of about £2,636 there were Oldknow's own debts, which amounted to £1,548, including a loan from A. Crompton, Esq., of £1,000.

In spite, however, of this favourable balance the problem mooted by Oldknow in his letter to S. & W. Salte in April of finding capital for expansion of his business had not yet been solved. It is true that both the London firms had given him permission to draw upon them, *i.e.* to borrow money on security of goods not yet delivered, and, if drafts made in round numbers represent such loans, the firm of S. & W. Salte advanced £670—their total payments for the year being £1,387. But such temporary advances, however helpful, did not enable Oldknow to realise his ambitions, and it is clear from later letters that Messrs. Salte were not prepared to invest larger sums for longer periods in the muslin business. The one man who was known to be rendering this kind of aid in other directions was Richard Arkwright, then the recognised leader of the cotton industry, and he likewise had in his hands the main supply of one of Oldknow's chief materials.

It is not surprising therefore to find the book of Sundry Expenses recording two visits of several days each to Derby and Cromford in November and December 1783, and three longer journeys, one of them lasting a fortnight, to Cromford and Nottingham in January and February 1784. These long discussions with Arkwright led to business arrangements including a loan of £3,000 at 5 per cent., which was to enable Oldknow to start his manufacture on a larger scale at Stockport.

"CALVERTON, *Feb.* 16, 1784.

"DR. BRO,

"You will naturally suppose my thoughts have been considerably employed in thinking of this new employment but whether a Bleacher or Spinner I am as yet undecided. The profits Mr. A. seemed to think equally advantageous, though Bleaching in all probability the most lasting. What Mr. Arkwright says with respect to the profits of Bleaching may be true, but cannot say my throat is sufficiently wide to swallow the information for fact. I don't know whether I told you what he said. At least you had not time to consider it properly. I shall therefore repeat it for your better consideration and desire you to send your observations, as I am certain a part of it you must understand He said if I return £1200 per annum by Bleaching I should not have above 300£ to pay out of it. He further said if you would manufacture 200 pieces per week the Bleaching and Dressing of them would be 2/– per piece that would amount to £20 per week and that I should clear £12 out of the 20£. These are profits in my opinion too great for the returns, so great that every manufacturer would be his own Bleacher; however I shall be well satisfied if the profits are half so much. Now the idea I have formed of the Spinning I should prefer for two reasons as the profits are said to be nearly equal. The first is if I was fortunate there is a greater probability of extending the business in spinning, than Bleaching. The second and greater reason is that in the spinning I should live in or near a town where (probably) I should have the benefit of having my children educated better and at a less expence and have them more with me than I possibly can at an Odd Place. I dont know that I can advance anything more on this subject at present than to desire you to give them due consideration and inform me the result of your thoughts and enquiries if time permits. I should wish to hear from you as soon as possible for as soon as the weather permits I intend to visit Stockport but not before I receive a letter from you. I should wish to know the days you are at Stockport as in all probability I shall not go further."

"Febr. 17. We were extremely sorry to hear this day by uncle John, of the very narrow escape you have had of being killed and we think it a great mercy that it pleased God so miraculously to deliver you by the turning of the saddle. I shall go purposely to Nottingham tomorrow where I expect they will have had another account; from whence I shall send this."

"Feb. 18. I have been this day to Nottm in hopes to have heard some further account of you but am sorry to say I have been disappointed but hope the old saying of no News is good News may be verified in you. and hope by this time you are as much recovered as circumstances will admit. I can say that all your friends are extremely sorry for your misfortune and very anxious for your recovery. Pray let us hear from you soon. My wife joins me in respects to all friends and in sincere wishes for your speedy recovery.

"from your affectionate Bro, THOS. OLDKNOW.

"N.B.—The Bond that I expect to execute on Saturday in your Behalf, I believe uncle T. wishes to see you at

Cromford in order to have things quite complete when as soon as you are able to perform such a journey I think a good Scheme. But his sentiments you no doubt know before this."

S. & W. Salte to S.O. :

"LONDON, *Jan.* 8, 1784.

"We are very unequal to the task of advising you with respect to your accepting or not accepting Mr. Arkwright's proposals, they have more generosity in them than we [expected]. You are so much more capable of judging of your own disposition & situation than we are that the matter must rest with yourself—When you have balanced the conveniences & inconveniences of this offer you will decide upon it & leave the success of it to Providence. If you can borrow money to enable you to go on by yourself we think with you that the prospect will be best, if not, & you can really make such terms with Arkwright as you mention would have you close with him. You will take all special care in your Articles, so as to have nothing to repent of hereafter. We wish you every kind of prosperity and happiness in either situation—to engage in a Manufacture ourselves we think wd be improper & indeed all our money is so much engaged in the printing Muslin & Irish Trade that we find none to spare. Wherever you are situated or with who soever conected we shall give you every proof of our Friendship & regard. We shall be glad to hear your determination."

The details of the bargain have not been preserved, but its broad significance is clear enough. The new inventions had endowed certain forms of fixed capital with unexampled powers of production, and the disposal of the surplus thus created gave to the possessors of that capital the initiative and direction of new industrial enterprize. Although for some years to come Salte remained Oldknow's chief patron and adviser, it was the influence of Arkwright that from this time onwards determined his career and gradually transformed him from an "eminent manufacturer" into a "captain of industry."

Chapter II.

THE INDUSTRIAL REVOLUTION AT STOCKPORT

TO one born in a cotton town a cotton factory is not a factory nor a works, but a mill. There are historic reasons for this. The first generation of the factory system was an age not of steam but of water Under the stimulus of the new inventions, two ancient devices, the canal and the water wheel, were born again into a higher sphere of usefulness. Within fifteen years of Arkwright's venture at Cromford the glens of Scotland and the valleys of Wales, as well as every considerable stream on both sides of the Pennines, were being turned to industrial uses. But long before the jenny or the water-frame was heard of, the silk-mills of Derby and Stockport were being worked by water-power, and the story of the Stockport development affords a unique illustration of the transition from mediaeval to modern conditions of industrial enterprize.

Readers of Messrs. Bennett and Elton's *History of Corn Milling* will learn that at the time of the Norman Conquest, when mills were the sole and widely diffused form of industrial capital and when mill-shares were almost as common as they are to-day, the country between Ribble and Mersey was the one part of England where property of this kind was absolutely unrecorded. Later on there were plenty of manorial mills belonging to the king and to private lords in South Lancashire, and it was here and in the West Riding that the rapid growth of the population produced the greatest increase in the profits of the monopoly enjoyed by the mills and made the burden correspondingly grievous. After riots in which a mill was burnt, and four inhabitants shot, Manchester in 1758 got an Act of Parliament removing the monopoly as far

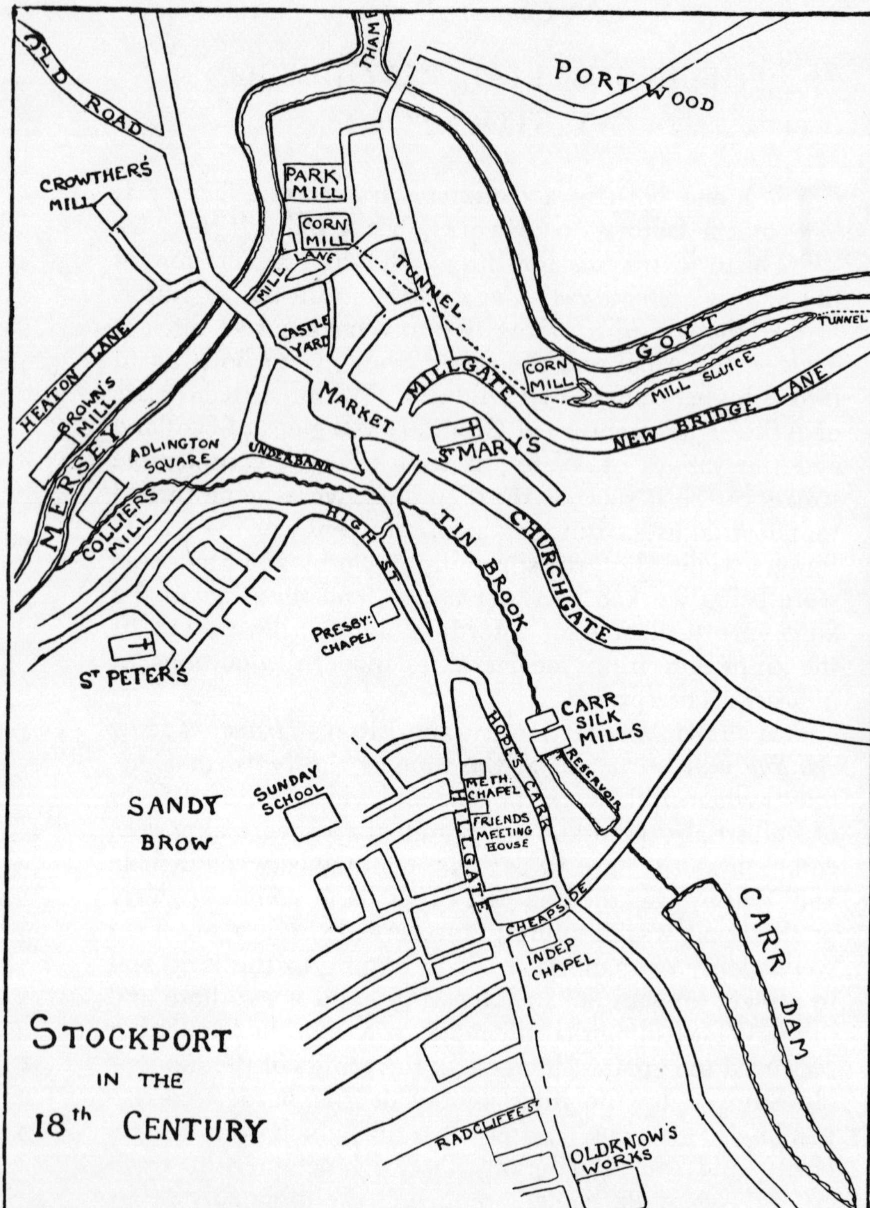

OLD ROAD
THAME
PORTWOOD
CROWTHER'S MILL
PARK MILL
CORN MILL
MILL LANE
TUNNEL
CASTLE YARD
HEATON LANE
BROWN'S MILL
MERSEY
MARKET
MILLGATE
CORN MILL
GOYT
TUNNEL
MILL SLUICE
NEW BRIDGE LANE
ST MARY'S
ADLINGTON SQUARE
UNDERBANK
COLLIERS MILL
HIGH ST
TIN BROOK
CHURCHGATE
PRESBY: CHAPEL
ST PETER'S
CARR SILK MILLS
HOPE'S CARR
RESERVOIR
METH: CHAPEL
FRIENDS MEETING HOUSE
SUNDAY SCHOOL
SANDY BROW
HILLGATE
CHEAPSIDE
INDEP CHAPEL
CARR DAM
RADCLIFFE'S
OLDKNOW'S WORKS
STOCKPORT
IN THE
18th CENTURY

as wheat was concerned, but the toll on malt continued to furnish for another century the chief revenue of the Grammar School.[1]

At Stockport the Warren family derived a considerable part of their manorial revenue from this source. In 1644–5, when their property was in the hands of sequestrators appointed by Parliament, the corn mills on the Mersey produced £49 and the Little Brook mill £33 out of a total of £285. The former of these lay under the Castle Hill and across the neck of the peninsula called the Park, which, along with the market place on the hill top, formed the centre of the town. The water-power had been improved by the construction of a tunnel cutting across an extensive winding of the river, and at the beginning of the 18th century it had already been put to other industrial uses. By the side of the corn mill there stood in 1732 a logwood mill erected by the lord of the manor for the use of the manufacturers who were now his principal tenants. That year marks an epoch in the rise of the Factory System. The patent granted to Sir Thomas Lombe for silk-throwing machinery had just expired, and Parliament, instead of renewing it, bought the model for £14,000 with a view to encouraging the extension of the industry. Stockport was the first town to take advantage of the opportunity. A partnership of six members was formed consisting of three Stockport gentlemen, one of whom was a member of the Warren family, two chapmen of Heaton Norris, and a silk merchant of London.[2] The lord of the manor, Edward Warren, Esquire, agreed to grant a piece of land next to his two mills and the use of his water-

[1] R. Bennett and John Elton : *History of Corn Milling*, Vol. II., pp. 105, 164 ; *Ibid.* III., pp. 207, 257, 270–82.

[2] Heginbotham : *Hist. of Stockport* II., 318.

power for a small chief rent and five per cent. of the profits.[1]

The original intention of the partnership had been to hire out the mill engines, when erected, to manufacturers, including perhaps some of its own members, but this proved impracticable and the proprietors found themselves obliged to undertake the business at their own "risque and hazard." It is significant of the relations between industrial and commercial capital at this period that whilst the mill and machinery only cost £3,000, a further capital of £9,000 was required to carry on the silk trade. The total of £12,000 was divided into twelve shares of £1,000 each.

On this basis a new start was made in 1736. In view of the change in the character of the enterprize, the lord of the manor was to commute his claim for five per cent. of the profits for a payment of £10 a year. At the same time he is said to have become the proprietor of three of the twelve shares and to have sold them for £3,000 to the leading London partner, who was however to retain the money as security that a lease for a thousand years which had been promised would be duly granted. Soon afterwards the lord of the manor died suddenly and a private Act of Parliament had to be procured to authorize the granting of the lease by the widow, who thereupon received the £3,000. It would appear from these transactions that the three shares were handed over in consideration of the promised lease, and that a quarter of the capital of the partnership was represented in the land and water rights secured, the chief rent in addition to the £10

[1] The account of the transference of ownership of the Park Mills given here and in a later chapter is based upon investigation of the title deeds and other legal documents which Messrs. Johnson of Henry Marsland Ltd. very kindly placed at my disposal.

already mentioned being only £2 5*s.* 9*d.* a year. It should however be added that a lease of the logwood mill was simultaneously granted to the partners for 60 years at a rental of £20 2*s.* 4*d.*

The partnership thus founded was in effect a seventeenth-century joint-stock company without a charter. It had the same mixture in its shareholders of landed gentry, distant mercantile capitalists, and managing partners on the spot, the same expansible shares tending to be alienated and split up, and the same inefficiency and instability for the purposes of business. When the partnership was dissolved in 1768 only one of the seven mercantile firms then interested bore the name of an original shareholder, whilst the other five shares were represented by thirteen persons, the legatees and executors of previous shareholders, amongst whom were two clergymen, a surgeon, a widow, and two spinsters. The one redeeming merit of this cumbrous arrangement was that it facilitated the co-operation of the various interests concerned in the enterprize—the manufacturer, the country gentleman, and the London merchant—and provided them with capital at a time when the means of communication and the machinery of credit were comparatively undeveloped. The wonder is that the firm, with whatever modifications, should have lasted so long. It had soon to face competition. As early as 1744 a rival body of capitalists obtained authority from Parliament to dam up the little stream called Tin Brook which entered the Mersey through the cleft between the Castle Hill and the larger hill upon which the extension of the town was then taking place, and up the side of which, parallel to the Tin Brook, ran the Hillgate, its main thoroughfare and the chief road from Manchester to London. The valley of the stream was known as the Carrs, and the Carrs silk mill thus erected below Middle

Hillgate became the second important starting point of the Factory System in Stockport. A third locality of historic interest in this connection is Adlington Square, which lay near the junction of the Tin Brook with the Mersey. Here a silk mill was built which may probably be identified with the one afterwards taken over by William Radcliffe and Thomas Ross.

Before 1768 half-a-dozen silk mills were at work in the town. In the edition of Defoe's *Tour* published in 1769 Stockport is described as

> a large and handsome town occupying three hills and three valleys which are so serpentine as to form many pleasing prospects of churches, pieces of water, etc., with the large silk mills belonging to the chief tradesmen of the place. It is inhabited by a great number of gentry and well filled with warehousemen who carry on the check, mohair, button, and hat manufactures. Here the raw silk is chiefly thrown and prepared for the Spitalfields weavers by six engines the buildings of which are of prodigious bulk, one of them containing above 45,000 movements which fill the spacious room up to the fifth storey and are all put in motion by one wheel that goes by water. . . . At this place poverty is not much felt except by those who are idle, for all persons capable of tying knots may find work in the silk mills, which employ near 2,000 people and where children of six years old earn a shilling a week and more as they grow capable of deserving it.[1]

[1] Defoe: *Tour*, edition of 1769, II. 397. The schedule attached to the Indenture transferring the Park silk mills in 1768 describes the machinery as follows: In the lower and higher winding rooms 18 double engines, each engine employs 128 spindles and the like quantity of swifts and bobbins. . . . In the lower and upper mill rooms spinning mills each 576 spindles, and 3 toter mills each having 384 reels in the doubling room, 140 wheels with Jacks and glass eyes. In the Dutch Mills, 4 double Dutch mills, each 160 spindles, etc.

From this broad survey of the rise and progress of the silk industry in Stockport we may turn, for the purpose of tracing its decline, to the history of the only firm of which we have details. The company or partnership operating the Park and the Logwood mills disposed of them in 1768 for £2,120 to Richard Blackburn of Wood Street, London, silk merchant, who thenceforward became a Stockport silk throwster. He took as a partner Thomas Tatlock, silk broker, but the firm became bankrupt in 1773. Tatlock then withdrew from the business, but he was able to arrange on Blackburn's behalf for the re-purchase of the mills from the creditors for £2,620, and for loans on mortgage from two firms of Huguenots in London—Dalbiac & Co., silk weavers of Spital Square, and Paul Agutter, warehouseman of Crown Court, Cheapside—Tatlock himself, the third and principal creditor, being described as a silk merchant of Wood Street, London. After a struggle for ten more years, during which he paid off half his debts, Richard Blackburn sold the mills and the remainder of the lease for £3,500, exactly half of which went to satisfy his three creditors. The new purchaser was Henry Marsland, who had hitherto carried on cotton spinning and manufacturing at Bosden about three miles outside Stockport to the south, and who, from the extent and the success of his subsequent achievements, may perhaps be regarded as the father of the cotton industry in Stockport.

Henry Marsland took over the Park silk mill to adapt it to cotton spinning a few months before Samuel Oldknow set up his manufacture at Stockport. At that moment Stockport had already passed through one phase of the Industrial Revolution and was just about to enter on another. By the time that the new spinning inventions began to be widely adopted the whole framework of the Factory System had already been erected at Stockport.

Not only water-power and buildings, but also the available child labour could be readily transferred from silk to cotton. These however were only the more external or the less essential of the conditions required for the new industrial order. Of the more creative factors three may be specially noted. One of these was the organizing ability displayed by captains of industry like Oldknow and Marsland. A second was inventive capacity, not only in its rarer and more conspicuous forms, but also of the sort exhibited in the resource and ingenuity of the skilled mechanic. And a third was the higher degree of social mobility shown in the emergence of new classes and of new modes of relationship between classes embodied in new forms of association. All these factors in creative social evolution were already operative at Stockport when Oldknow set up business there. Of the first sufficient evidence has already been given, and a word or two may be devoted to the second and third.

The story of industrial invention at Stockport finds its starting-point in the achievements of Lawrence Earnshaw, of Mottram, who got his only training from a Stockport mechanic during the earlier part of the silk mill period. When a boy, we are told (by Aikin), he

> was put apprentice to a taylor and afterwards to a clothier; but neither of those employments suiting his genius, after serving both for eleven years, he put himself for a short time to a clock-maker, one Shepley of Stockport. By the force of native abilities, with the very little instruction such an education could give him, he made himself one of the most universal mechanists and artists ever heard of. He could have taken wool from the sheeps backs, manufactured it into cloth, made that cloth into cloaths and made every instrument necessary for the clipping, carding, spinning, reeling, weaving, fulling, and dressing, and making it up for wear with his own hands. He was an engraver, painter, and gilder; he could stain glass and foil mirrors; was

a blacksmith, whitesmith, copper-smith, gunsmith, bellfounder and coffin-maker ; made and erected sundials ; mended fiddles ; repaired, tuned and played upon and taught the harpsichord and virginals ; made and mended organs and optical instruments ; . . . and he carried so far his theory and practice of clockwork as to be the inventor of a very curious astronomical and geographical machine, containing a celestial and terrestrial globe to which different movements were given representing the diurnal and annual motions of the earth, the position of the moon and stars, the sun's place in the ecliptic, etc., all with the greatest correctness. . . . He likewise, about 1753, invented a machine to spin and reel cotton at one operation, which he showed to his neighbours and then destroyed through the generous but mistaken apprehension that it might take bread from the mouths of the poor. . . . He also contrived a simple and ingenious piece of mechanism for raising water from a coal-mine. He was acquainted with that equally self-taught genius the celebrated Brindley and when they occasionally met they did not soon part.

Whatever allowance we may be disposed to make for local hero-worship in this recital of marvels, there will remain enough to justify us in assuming that the Italian machinery for silk throwing with its 45,000 motions provided a stimulus to which the inventive genius of the neighbourhood did not fail to respond. Earnshaw left behind in the district not only a tradition, but successors. Owing perhaps to the later partial decay of the cotton industry at Stockport, the part played by the town in the history of textile invention has not met with full recognition. It was at Stockport in the first decade of the 19th century that Johnson, Radcliffe, and Horrocks did for power-loom weaving what Arkwright did for power spinning, though with results less remunerative to themselves. But twenty years before this, when the

[1] J. Aikin: *A Description of the country from thirty to forty miles round Manchester*, pp. 466–8.

future of Arkwright's patents was still undecided and before Crompton's mule had been widely adopted, inventors were busy at Stockport devising new improvements in carding and roving machinery. Professor Daniels has drawn a lively picture of the successful resistance offered in Manchester to Arkwright's attempt to recover his patent (granted in 1775 and annulled in 1781) for carding, drawing, and roving machines, to consolidate it with his earlier patent for roller spinning which was to expire in 1783, and to get both patents continued till 1789. The effect of such a grant would have been to give Arkwright legal control of the mule which was covered by the patent for roller-spinning, and to have made the jenny-spinners dependent upon him for the use of carding and roving machinery.[1] The Manchester Committee of Trade did not confine itself to Parliamentary opposition, but proceeded to collect £200 to reward an inventor of new roving machinery. A fortnight after the announcement of this in the *Manchester Mercury*, John Swindells, an engine maker, of Stockport, informs the public in the columns of the same newspaper (April 30th, 1782) that "he makes machines that will card and rove cotton at one and the same time on an improved plan . . . with a callender feeder" and adds the testimony of Mr. Josiah Whittaker, of Stockport, who "has made a trial of roving from the above Engine upon a common Jenny and has found it answer beyond his expectations as only one thread broke down in the space of one hour upon thirty spindles and made far superior yarn than the hand-slubbing did upon the same Jenny."

In Wheeler's *Manchester Chronicle* for October 19th, 1782, an old man retired from business writes to express his "pleasure and wonder at the powers of machinery

[1] Daniels, p. 105.

in the Cotton Branch." "In my juvenal days," he says, "when cotton was carded by hand and spun upon the spindle" . . . a pound a day was esteemed good working . . . but see the contrast . . . " "At Mr. James Gosling's factory in Stockport I saw a double carding engine made by the ingenious Mr. Swindale which for expediting business surpasses all I ever saw; it carded last week 520lbs. weight of cotton in five days." As doubts were cast by malicious and ill-natured persons upon the truth and genuineness of this unsolicited testimonial, John Swindells proceeded to insert in the *Manchester Mercury* of October 29th, 1782, unsolicited affidavits to the same effect by the carder who did the work and others at Gosling's factory, which, it is interesting to note, was situated in the Corn Mills. At the close of this advertisement "J. S. begs leave to acquaint the Public that he continues to make machines that will card and rove at the same time."

John Swindells was apparently not the only maker of these machines in Stockport. *The Manchester Mercury* for February 18th, 1783, which prints an appeal of the Manchester Committee to all cotton manufacturers to unite in opposition to Arkwright's petition, contains also an advertisement by William Mycock, of Stockport, of machines (approved by the Manchester Committee of Trade) "which card and rove cotton also wind it upon spindles at the same time in one operation. The roving and slubbing from these machines will spin upon the jenny as well as that done by hand, as it is not fined or drawn upon rollers."

Now, the invention that played the most important part in solving the problems of the years 1783–6 was the billy, and this has always been attributed to an unknown Stockport man who is said to have had a premium voted to him by the grateful jenny-spinners of the district.

Up to this time [says Professor Daniels] the mule had been used solely for the spinning of yarn. The rovings for spinning had to be made either on the spinning wheel or by Arkwright's machinery. The "Billy" was a modification of the mule, or rather a combination of the mule and the jenny; but instead of spinning rovings into yarn it made carded cotton into rovings. With this machine rovings could be made for the use of the mule, the jenny, or even the water-frame, to any required degree of fineness, and at a greatly reduced cost.[1]

The "Billy" as thus described and as depicted in Ure's *Cotton Manufacture* differs from the new machine advertized by Swindells in not including the process of carding, and it is not difficult to account for the separation of this process in view of the patent rights claimed by Arkwright over carding machinery, which were reaffirmed for a short time in 1785. It seems, therefore, not unlikely that the hitherto unknown inventor of the billy was John Swindells of Stockport, and the fact that Swindells was declared a bankrupt in 1786 will tend, in the minds of those familiar with the history of invention, to strengthen rather than to weaken his claim. Perhaps his business troubles furnished the occasion of the rally of the grateful jenny-spinners on his behalf.

With these mechanical developments the growth of a separate class of organized wage-earners is closely connected. Whilst the invention of the billy prolonged the usefulness of the jenny, and the smaller jennies and mules continued to be used by cottage spinners, the prevailing tendency was towards the increase in size of both jennies and mules and towards the collection of them in small factories where they were operated in conjunction with carding and roving machinery by workers who did not own them. The earliest definite evidence hitherto published of the organization of these

[1] Daniels, pp. 123–4.

journeymen spinners is to be found in the articles and regulations drawn up by the Friendly Associated Cotton Spinners of Manchester at the Three Horse Shoes in the Market Place, 1795, and a similar club is said to have existed at Stockport as early as 1792. An advertisement, however, in the *Manchester Mercury* of June 21st, 1785, not only proves that the Stockport society was in existence at that date, but casts such a full and interesting light on the beginnings of trade unionism in the spinning industry that it is here given in full.

The Committee now acting for the Friendly Society of the Cotton Spinners of the Town of Stockport request that all Spinners of Cotton who shall be out of employ do immediately seek out for work and get employ as soon as possible in any Branch, Trade or Calling whatsoever but not to work under the usual Prices; and it is further desired by the Committee that every person who have left their Trade and Calling do immediately return to the same as soon as possible, as the Committee now acting will not relieve such Person on any Account whatever who refuse to comply after this public notice being given, and it is further desired by the Committee that no Person or Persons whatsoever who is employed in any Part of the Cotton Business do make or cause to be made any disturbance or Riot on any Account whatever & no person will give any money to any person but the Committee as every one in the Committee will have a paper stamped in similar manner that the Public need not be imposed on by any person. Therefore all well disposed Persons who are inclined to subscribe any small Trifle to the Committee who shall produce the Paper it will be gratefully acknowledged.

N.B.—No person to give to any of the Committee without paper & that there shall be two or three of the Committee together to receive any small Trifle but not to deliver Money to one Person alone.

May an honest heart never feel distress. [Eleven names appended.]

There is nothing in this notice to prove that the society of Stockport spinners was in 1785 entirely composed of journeymen, and it is quite possible that it included domestic workers who owned jennies. Amongst the check-weavers and smallware-weavers of Manchester small masters and journeymen were, as Professor Daniels has shown, combined in one organization before the middle of the 18th century for the defence of what was practically a wage-earning status, and in spite of prosecution and their disavowal of combination in 1759–60 the smallware-weavers were found in 1781 carrying on collective bargaining with the manufacturers who employed them.[1] The earlier history of trade unionism is mainly concerned with a composite class interest of this kind, and this phase in the case of the handloom-weavers lasted till well on into the 19th century.

For this reason the body of legislation which represents the class interests of the employers in the textile industries is not directed merely to the prevention of combinations amongst wage-earning small masters and journeymen, but also to the protection of circulating capital entrusted to domestic workers. The Worsted Committees authorized by the Act of 1777 for the West Riding, Lancashire and Cheshire, and by later Acts of 1784, 1785, and 1790 for Suffolk, for the Midland Counties, and for Norfolk, armed the combined employers, who were unwilling to expose themselves singly to the resentment of the spinners and workpeople, with special powers for the detection and punishment of those who embezzled their materials.[2]

[1] Daniels, *op. cit.*, pp. 40–53. Prof. Daniels informs me that the check weavers in 1768 were insisting on the silk weavers joining their association, and that in 1786 there were three silk weavers prosecuted for combination.

[2] Heaton: *Yorkshire Woollen and Worsted Industries*, p. 420 *et seq.*

The inspectors appointed by these committees could procure a warrant from the justices to search private houses, and if any suspicious goods were found the accused person was deemed guilty till he proved himself innocent. By the first of these Acts the period for which a domestic worker was allowed to retain the material given out to him was limited to eight days, and if he did not return it properly worked up within that time he was liable to the same penalties as for embezzlement. In the interval his employer could enter his workshop to inspect his operations. There was not much left of the independence of the small master except the choice of his hours of labour, and against this freedom he had to set off the continual risk of unemployment. His remuneration was frequently lower and his domestic privacy less than that of the wage-earner in a factory, and he was still liable to some of the risks and disabilities attaching to the possession of industrial capital.

But there was another series of statutes enacted during the same period and inspired by the same interests whose significance does not appear to have been noticed. The more independent type of small master was not extinct, and his relation to the employing class was frequently one not of nominal but of real indebtedness. He bought his materials on credit, paid interest on the debt, and in default of payment had to be sued for it in the ordinary courts. What was now being demanded and obtained from Parliament was the erection of special Small Debt Courts in industrial areas in which the larger industrial and mercantile capitalists might sit in judgment on their own disputes with smaller capitalists and secure a speedier process against them. Such courts were set up in the West Riding in the same year as the Worsted Committee (1777), with two hundred and thirty-seven assessors appointed locally and jurisdiction over nearly a score of parishes and townships.[1]

Stockport enjoyed the advantages of the special legislation against embezzlement, but found it inadequate. It is true that we hear in 1789 of a "person" being publicly whipped in the pillory for this offence after conviction before the Rector of Stockport.[2] But in May 1787 we find Holland Watson, another of the Stockport justices, explaining to Oldknow that the law requires two magistrates to sign a search warrant against a suspected person. A single justice, he adds, may issue a summons against a spinner not returning the wool worked up in due time (which is the common mode of proceeding), but two justices will be required to convict. "This I think a great inconvenience in a manufacturing country like this, where there are few magistrates."

A month later Oldknow received the following letter from the wife of a spinner in Mellor:

"SIR,

"I am sorry to inform you that your Cotton is spun & sold & until I received your Letter I did not know but my Husband had delivered it to the right Owner, having seen my Husband sinse, who is exceeding sorry for what he has done (being then in great need of Money) humbly desires the Favour not to be put to any Expence about the recovery of the same and in a very short time he will either pay you the Value in Cash or buy and deliver to you the same Quantity, of equal or better Quality either in Wool or Yarn at your Option. Your compliance and Answer to the above will greatly oblige her and him who are, Sir,

"Your obliged Servts

"EDWARD }
"RACHEL } BOWDEN."

[1] 17 Geo. III. cxv.

[2] Wheeler's *Manchester Chronicle*, Aug. 8, 1789.

In view of these facts it is not surprising that the manufacturers of Stockport had already attempted to establish a Small Debt Court. This was part of a larger scheme of municipal reconstruction for which in 1785 they petitioned Parliament.[1] Stockport was an ancient borough with a charter and a mayor, but the real power was in the hands of the Steward of the Lord of the Manor, who selected the jury of the Court Leet, paid the mayor's salary, and chose the constables. After half a century of vigorous industrial growth the town's reformers thought that the time had come to put an end to this seignorial régime. They proposed that a body of commissioners should be appointed consisting of the Lord of the Manor, the Rector and the Vicar of the two churches, and all residents with £600 of personal property, who should enclose and sell the common and waste, devote the proceeds to the improvement of the town, assume the chief functions of municipal government, and act as a court for the more speedy recovery of small debts.[2] This programme was too Utopian for immediate success. It took forty more years and three successive local Acts to carry it out. But, apart from its interest as a landmark in the growth of public opinion, the reasons given for the establishment of a Small Debt Court afford valuable evidence of the stage that had been reached in the town's industrial development.

Within the said town [says the petition] there are several considerable manufactures as the throwing of silk, woolcombing,

[1] S. and B. Webb: *English Local Government.* Manor and Borough I., 169–70. Dr. F. H. Spencer, in his *Municipal Origins*, has shown to what a large extent the work of the municipal reformers of 1835 had been anticipated by local Acts during the latter half of the 18th century.

[2] *House of Commons Journal* 1785, p. 583.

making of hats, buttons and twist, and spinning, twisting, weaving and manufacturing of cotton and yarn and printing of linen and cotton goods which employ great numbers of people who contract many small debts amounting yearly to a great sum of money, and though they are well able, refuse to pay such debts, presuming on the difficulties their creditors will lie under.

On the whole we are perhaps justified in concluding that in 1786 the flourishing industry of Stockport was based to a greater extent on the old order still represented by Samuel Oldknow than on the new order represented by Henry Marsland. Tunnicliffe's *Topographical Survey* (dated 1789, but not sufficiently up to date to mention either of these important new comers) gives a list of some thirty manufacturers and merchants in the town, only two of whom were said to be connected with the silk industry, three were button and mohair manufacturers, whilst the rest were designated as check (five), calico (thirteen), or cotton (ten) manufacturers. The term "manufacturer" had hitherto been applied sometimes to the domestic weaver but more commonly to his employer. In the 19th century it was transferred to the employer in any industry under the factory system, although in Lancashire it has been appropriated by the employer in power-loom weaving as distinct from the master spinner. During the transitional period with which we are now concerned the term was often applied to the small or large factory master who was carrying on carding and roving by water-power, was beginning to collect the jenny-spinners into his mill, and who not infrequently gave out yarn to weavers of calico or other cotton fabrics. It would be therefore not unnatural to regard the ten "cotton manufacturers" of Tunnicliffe's *Survey* as the representatives of the new form of enterprize, but there is no evidence to confirm this identification, whilst on the other hand we know that of the thirteen calico manufacturers five at least were already or shortly afterwards became master spinners. One of these, Thomas Brentnall, supplied yarn to Oldknow, and we have a detailed account of their business relations for over two years. The other four—James Brown, John Collier, Joseph Heaward, and John Whitaker—were soon to become the chief rivals of Henry Marsland in water-power spinning on a large scale.

These names suggest a large topic that must be touched upon, however briefly, before the close of this chapter. On June 12th, 1784, John Wesley "found a lovely congregation at Stockport much alive to God." In the trust deeds dated a month later of the larger chapel that had been erected on the old site in the Hillgate occur the names of John Whitaker, Joseph Heaward, and John Collier, as well as that of Jeremiah Bury,[1] check manufacturer, who later on became a leading spinner. The new social mobility which was a main factor in the economic development of Stockport was intimately connected with the emergence of new forms of religious community. At Duckenfield in the parish of Stockport was formed one of the earliest Independent churches, and here too, according to Sewel, was the first beginning of George Fox's preaching. The Quakers found sympathizers during the time of persecution amongst the poor of the town, and one of their society whose goods had been distrained in 1685 took office as Mayor in 1721.[2] In the same year the movement towards a more liberal theology induced a large body of the Independents to forsake the ministry of a high Calvinist in the Tabernacle Chapel and to set up a Presbyterian church in the High Street.[3]

[1] Heginbotham, II. 67.

[2] Heginbotham, II. 13. W. Sewel: *Hist. f the Society of Friends*, p. 36.

[3] Heginbotham, II. 57–9.

The spirit of enquiry infected the earliest disciples of Wesley at Stockport and led them to assert the "perseverance of the saints" against his doctrine of "sinless perfection." In healing this schism a leading part was taken by John Oliver, the son of a Stockport shopkeeper, whose spiritual experiences at the age of sixteen recall, alike in their exaltation and their extravagance, the stories of the *Acta Sanctorum.* Oliver had been convicted of sin, had found salvation, had again fallen from grace, and twice attempted suicide before he was twenty.[1] The year of his conversion (1748) saw the publication not only of John and Charles Wesley's *Hymns* but also of Hume's *Essay on Miracles* and of Montesquieu's *Esprit des Lois.* Between the inner world of the Methodists and that of the philosophers there might seem nothing in common, but it would be hard to say whether the "enthusiasts" or the enlightened humanitarians contributed most to raise the new social structure which was the positive outcome of the Industrial Revolution.

The Sunday School movement may not unjustly be attributed to a conjunction of both these influences, though it was the religious enthusiasts who did most of the work.[2] Started by the benevolent proprietor of a newspaper at Gloucester in 1780, it received the blessing both of John Wesley and of Adam Smith and aimed at the secular instruction as well as the religious edification of the young. At Stockport the movement was started in 1784 by a combined effort of the churches. Eleven schools were provided in which twenty-three teachers, who received eighteen-pence each for the day, taught some seven or eight hundred children from nine till six with an interval for dinner and another for attendance at evening service. After about ten years of this joint effort

[1] Southey: *Life of Wesley*, Chap. XVII.

[2] Findlay: *The Children of England*, pp. 111, 120–1.

the management of the schools was gradually taken over by the several churches concerned, which were thus enabled to call forth and dedicate to the work that great wealth of personal devotion which is the central fact of Sunday School history.

For great as the educational service of the Sunday Schools was—and for two generations the new working classes had scarcely any other means of secular instruction—their social service was greater. In the new factory towns, amidst the social degradation and anarchy produced by violent economic change prolonged through twenty years of war, they were the sole organs of a community that transcended the fierce antagonism of misconceived class interests. In them the masters, foremen, and workers of the factory met on the common ground of mutual service and here the industrial democracy depicted by the dramatist in *Hindle Wakes* came silently into being. Of this social significance of the Sunday School Stockport furnishes a unique illustration. When the churches set up their separate schools the undenominational nucleus survived in the Stockport Sunday School, which, unsupported by any church organization, and strong only in the social cohesion and in the tradition of loyalty and service which it had developed within itself, has retained its position as one of the leading institutions of the town and still claims to be the largest Sunday School in the world.[1]

[1] Heginbotham: *History of Stockport* I. 348–50, II. 74–5, 385–98; W. I. Wild: *History of the Stockport Sunday School.*

Chapter III.

THE HAND-LOOM WEAVERS AT STOCKPORT AND ANDERTON

(*By* G. Taylor, M.A.)

On July 21st, 1784, Oldknow rode over to Stockport taking with him a £50 banknote and £25 in bills which was half the purchase money for a house and land in the Hillgate belonging to Mr. Giles Walmsley, the other half of the price being payable within six months.

But the new warehouse had already been opened early in 1784, and a canvass made amongst the weavers of the surrounding villages by Oldknow's manager. His efforts resulted in about 90 weavers calling on February 3rd, 1784, at his Hillgate premises for supplies of raw materials. During this first month he gave out some 530 lengths of warp, with weft in proportion, and 90 reeds, probably amounting altogether to a sum little less than £1,000. In the same month he received back 270 cuts from 70 weavers, who earned among them £140. Of the 90 who had volunteered for work some 20 either failed to do the work satisfactorily or found it uncongenial and hence disappear from our notice. In later years many of these people returned to the loom and made efficient workers. By October of this first year over a hundred weavers had found a regular connection with Oldknow. A letter from the Stockport manager to Oldknow at Anderton shows us the undertaking in its earliest stages.

"*June* 14, 1784.

"Sir

"I Recd yours of the 11th Inst & Have Been at Mr. Bancroft But He has Set is House so I am Too Lat thear.

But so soon as it is Convenent I Desire you to forward My furniture as thear is a maney Things I am Whanting it will be Six weeks before Mr. Bancroft will Remove at the Soonest. I Have Been out amongest the Wevers & find it wd Not be Lost Time if I was to Go a few More Times. Thear is about 10 wevers Will Want Muslin Reed in the Corse of Next week. We have Non but one 60 6/4 Reed Emty. I desire you to forward Som Emadetly or we shall Not be able to serve them According to Promis—I think the 54 5/4 Callo will Not Go under 9/– p Cut for weving & then you May Have aney quantity and Good wevers. I Hope you will make your Selph Easey Consarning Mr Dale as He shall Not Eather see or Lern anething from Me While I am your Servant. We shall want a Little No 34 or 36 Twist for Strip. We are short of fine Weft I desire you to forward Som."

For the first few years none of the old fancy cottons, the fustians, checks and smallware, or goods with linen warps found a place beside calico and muslin. Calicoes were made in lengths varying from 21 to 31 yards of all widths between 3/4 to 6/4 yards. The two chief varieties were the blue check in which a dyed warp was used and the doosootee with a selvage in the middle. The essential feature of calicoes was that their patterns were printed in fancy colours by wooden blocks. They had long been familiar to the English public and early in the century had been the object of "protective" legislation. English workers had readily adopted the art of printing plain calico from India, but, as Professor Daniels[1] points out, there is no evidence of calico weaving or printing in Lancashire until after the middle of the century. Muslin was the generic term for the finest cotton goods. At this time they were made from yarn of counts between 50–70.

[1] Daniels, p. 22.

They were woven in all varieties of shapes and patterns, but were either 20 or 29 yards in length and were sold in half-lengths or "demies" after special attention from the bleachers, dyers, and calenderers. Manufactured in imitation of Indian goods, they received largely Indian names. Oldknow specialised in material for ladies' gowns, in Ballassore handkerchiefs, and in aprons. In the years 1784-5 his output was divided almost evenly between muslin and calico goods. All his hopes and ambitions were centred on the former, and the development of his business can be seen on the lines of their expansion and differentiation.

In spring 1784 they were represented by striped and checked varieties. In the first case part of the warp was coloured, in the second the weft was variegated also. In the summer plain muslins were apparently in greater demand and two or three weavers began immediately on the manufacture of the complicated figured goods, which necessitated the use of the draw loom or drawing engine.

The consumers of Oldknow's goods were mostly people of fashion. S. W. Salte indicated the classes of goods most suitable for this market. "Vary the spot Barleycorns, leaves, and other little fancy objects, in short an infinite variety may yet be made ; we do not despair of great attainments in this branch of trade. Ingenuity and patience and perseverance will yet work miracles." The weaving records show Oldknow's response to these requests. Fancy muslins, romals, cravats, cord and charr checks appeared in turn, until the catalogue of sales comprised a hundred varieties. Yet Oldknow was not unmindful of his calico weavers and always found a place for such articles. Calico, moreover, did not suffer much from the seasonal fluctuations which affected the muslin trade eight months in the year.

To add to their resources, Oldknow soon included shirting in his list (January 1785), and this shortly became an important branch. This cautious proceeding did not please the London agents, who were anxious "to take more risk for extra profitt." In increasingly strong terms they wrote requesting him to confine his efforts to the higher branches. "The Blue and White checks [calico] are quite a drag, indeed this trade is beneath your notice, ingenuity is not wanted ; leave them to Fowden and such fellows who have no reach of genius beyond them unless when they pirate and Copy from you." A final letter on this subject is more brusque in tone: "It is not satisfactory to us that you are on a level with others, we must be superior. Nothing but new things will please fashionable women, try your invention once more."

By this time (1786) Samuel Oldknow had obtained an assured position as the foremost muslin manufacturer. The expansion of his weaving connection had been continuous, and increases before the spring and autumn sales had become customary. The large expansion in 1785 may have been partly due to the removal of the taxes on cotton goods in April. Twelve months later the trade was encouraged by the decrease in the price of yarn that Arkwright was forced to make owing to the competition following the invasion of his roller-spinning patents.

By the autumn of 1786 Oldknow was employing over 300 trained weavers, who possessed among them at least 500 looms. The monthly wage bill was not less than £1,000, for 2,000 pieces of cloth. In addition to this quantity at Stockport, the Anderton weavers, who had abandoned the old fancy goods, were supplying about 1,000 pieces per month, and their number had expanded from 59 in 1783 to 159 in 1786. Altogether Oldknow's sales for November (1786) amounted to £6,000.

The weaving records, the ledger and journal, afford ample material for the study of the relations between Oldknow and his weavers. There is no evidence that Oldknow followed the custom of the district by employing "putters out." It is probable that several days a week weavers could be seen tramping along the roads leading to Stockport, with their bundles of cloth which they would leave at the Hillgate warehouse. A letter shows that some who lived at a distance from town journeyed up by waggon. At the warehouse they awaited their turn in the lobby, where forms were provided. In turn they passed into the "giving out room," where one or two assistants were engaged weighing out raw material. Here the weaver was supplied with a warp of four cuts length, and a suitable allowance of weft. The yarn would then be stowed away in the canvas bag brought for the purpose.

In a few cases raw cotton was supplied instead of weft. It appears in the ledger under the heading of "wool" (the 18th century term for raw cotton), and the weaver received extra payment for the spinning. At this date the weaver had long lost the power of working up the material as he liked, and was receiving a careful specification of the cloth required. Although every master weaver possessed at least one loom, it was customary for him to receive from the employer reeds and gears for each class of work. It would indeed have been beyond the means of his slender capital to keep a stock of these for all the numerous patterns he might be called upon to produce. Oldknow's stock of reeds and gears, valued at several hundred pounds, represented a considerable proportion of his assets and his staff of hired workers included several men to repair and make reeds and women to knit gears or healds. Figured muslins could only be produced on the newly invented and costly draw loom, or drawing engine. Oldknow had to provide these looms, which he let out on hire at the nominal rent of 6*d.* a week. It was a significant step towards the days when new inventions would take from the weaver his status as a small master.

On arriving home the weaver had to spend a considerable time preparing his loom, or "beaming." The warp, now in loose rope form, had to be threaded, by the aid of the weaver's hook, through a kind of metal comb called a raddle, a kind of reed with very wide dents and a detachable cap. The nature of the warp determined the pattern of the cloth. Some weavers had special patterns of warp reserved for them. In this pre-arranged order the warp was wound on to the beam, a roller 2 yards in length, which fitted at the back of the loom. The weaver's wife or apprentice turned the roller while he himself guided the threads.

The warp had then to be threaded through the leashes of the healds, which were suspended between the beam and reed, and finally through the reed itself. One end passed through each knitted leash, two through each dent of the reed. The latter, fitted in the swinging batten or lay, was composed of fine teeth of steel, brass, or cane fitting between two strips of wood.

After seeing that the warp and loom were in perfect alignment the weaver had finished his preparation, as the weft only needed threading from cops into the spool of the shuttle. Weaving remained at this period a man's work. Only one woman had an account with Oldknow, and she probably employed a journeyman. A cut of 20–30 yards on the average represented one man's work for a week. The drawing engine required the assistance of a boy, "a draw boy," to manipulate the ropes which regulated the gears. Plain woven fabrics such as calico and plain muslin were found by interlacing warp and weft threads alternately. The strength of this combination compensated for the lightness of the fabric. Figured cloths were obtained by passing one pick over two or more ends in the " twill weave." The difference was due to the heald action, which "shedded" the warp and so allowed the shuttle to pass between. Skill was required in this operation, for over-shedding caused unnecessary strain on the warp and under-shedding resulted in the end being broken by the shuttle. The fly shuttle, released by a peg, saved the weaver time and effort and it had been quickly followed by the adoption of the drop shuttle, whereby different shuttles with various wefts could be easily introduced. The final operation consisted in the beating up of the cloth, *i.e.* the newly woven part was tightened as the result of a sharp tap of the batten. The skill of the weaver was shown in his production of an even, faultless cloth.

As a piece was completed it was cut off the cloth beam and carried to the warehouse, where a room was reserved for receiving manufactured goods. In this " taking in room " the cloth was placed on the board, examined, weighed, and sealed. If passed as satisfactory the " cut " was placed in the drawer or shelves with which the room was provided, ready for despatch to the finishers. The receiver marked a weaver's ticket with the weaver's name and the goods returned together with the price in the current price list. This ticket was taken by the weaver to the counting-house, where he received his wages in exchange. The tickets were afterwards collected in daily bundles and checked.

Oldknow's accounts of weaving were based on a sound if cumbrous system of book-keeping. Every warp bore a number, so that damage or bad workmanship could be traced back to all who had handled the cloth. As the goods were returned particulars were entered in the " taking in book, " or journal.

Specimen Entry.

	Name.	Ledger No.	Cuts	Sort.	Price	
3 Feb.	Thos. Horsfield	614	2	56 5/4 Plain.	13/–	1. 6. 0.

The entries of this journal were posted twice a week into the weaver's account in the ledger. Although he no longer bought or received on credit his supply of raw material, the weaver was debited with the length of twist and the weight of weft he received and was credited for the quantity he returned in the cloth and remainder, together with the wage for weaving.

The remarks made in the column reserved for that purpose cast interesting side-lights on the relation of Oldknow or his representative and the men he employed. Apparently some of the more improvident weavers were in the habit of obtaining loans or advanced payments for their work. Amounts up to 10/– were commonly granted and a note made to that effect in the account. Occasionally a weaver returned a short length of cloth, and he was naturally open to the suspicion of cutting off a piece for his own use. These cases the manager left to the decision of Oldknow. The manager had the power to make allowances for bad warp or poor cops of weft and to deduct small sums for insufficient workmanship or faults. A curious item is the deduction of 1/– for a "mouse-hole."

With the growth of the business the manager became a highly important figure. The preservation of the Anderton correspondence reveals the duties of the manager who was left in charge at that place when, in 1784, Oldknow made his headquarters at Stockport, and who was responsible for the transactions with all spinners and weavers, arranging for the distribution of raw materials and instruments, and the return of manufactured goods. This involved a close watch on stock, accounts of which had to be returned weekly to Oldknow, through whom fresh supplies were obtained.

Clearly this post could be efficiently filled only by an occupant of practical experience in every branch of the cotton industry. A summary of his daily duties as suggested by Oldknow is added :

Monday	Packing day for weft and warp
Tuesday	Taking-in day for spinners
Wednesday	Taking-in day for weavers
Thursday	Packing day for cloth
Friday	Taking-in day for spinners
Saturday	Taking-in day for weavers

The Stockport warehouse was open four days per week for the return of goods by weavers.

The manager was responsible for paying the wages, and he checked the records. The prices of weaving were contained in periodical lists made out under Oldknow's direction, and lists for spinning and winding were sent in a smaller form. Every fortnight accounts were submitted to Stockport, and the Anderton manager received remittances of cash and bills. A letter shows him negotiating over payment in bills. The workers disliked this system, for they had to pay 3*d.* per £1 to the local shopkeeper for cashing. The manager undertook to cash bills with a banker and stop the premium at the rate of 22*s.* 8*d.* per £100 from their wages.

This account of the manager's functions will serve to explain the attempts of Oldknow's rivals to attract a man so intimately bound up with the business. At first sight, the remuneration for such onerous duties appears very inadequate. The salary " for servitude " commenced at

£35, with a weekly deduction of 5*s*. for board, and rose to £50. There are two explanations. The manager had either a business of his own or regarded his office as a preparation for the rôle of entrepreneur.

In this connection the records illustrate a significant aspect of the Industrial Revolution. It was from this new class of managers that the rising generation of entrepreneurs such as Radcliffe and Owen were recruited. There is clear proof that six of Oldknow's employees afterwards set up on their own account. Entering his service with a practical knowledge of the industrial branches, they quitted it with a working acquaintance of business methods, with the capacity to control men and organise a large-scale production. Swift, the Anderton manager in 1786, was a rival manufacturer to his former employer two years later. Generally, however, the new entrepreneurs remained dependent on the larger capital, credit, and superior market organization of Oldknow. Thus Horsfield became a yarn manufacturer, supplying Oldknow with raw material for manufacture. A clearer account can be given of Richard Heys. One of five brothers (three of whom were Oldknow's weavers) residing at Bullock Smithy, he was a member of the Stockport staff in 1785 and became a small master spinner in 1788. A record of his transactions with Oldknow some years later has been preserved. He purchased his supplies of raw material through the latter and distributed it among spinners of the district. At the period recorded he had added weaving to his sphere of operations. In January 1793 he received £79 for cord twist and £51 for calico. During the course of the year the spinning branch declined but the monthly payments for weaving rose to £87.

The following letters from Anderton illustrate the varied responsibilities of the manager there and his relations with his employer:

"ANDERTON, *July 4th*, 1786.

"SAMUEL OLDKNOW,

"I have herewith sent a list of the Weavers at Anderton as near as I can Only their is 5 ps of 5/4 Callo cut which I will get in as soon as possible——I have several 7/8 Callo. weavers waiting to know if they must have any more work—which are not in this list. I could wish to know if any alteration is to be made amongst the weavers —to let me know very soon as hay-time is coming on & if some more weavers gets work they are likely to hold it some time on that Acct—but I greatly expect Saml Oldknow at Anderton very soon if ever he can spare time to get so far. I am with due Respect

"THOS SWIFT."

"ANDERTON, *September 27th*, 1788.

"SIR

"These two weavers Jno Heys & James Wrostern have been to Enquire for work—but on finding we had got the number we was to have at present the[y] immediately determined to come to see you—if they meet with success I think you will have more of them. I sent Cart yesterday to tol Bar & there was neither any more Reeds nor Cotton. There was then near 20 Spinners waiting for it—how we shall get this day over I dont know. I desire you will let me know when we may depend on having some & also the Reeds. Weavers comes every day & I cannot tel what to say to them

"I Ham Yr Obt Servt

"T. HORSFIELD."

"P.S.—I wrote to you on Monday to let you know we was without Cotton

"Mr. Sam Oldknow Stockport."

"ANDERTON, *October 20th*, 1788.

"SIR

"I have altered the Spinners List of prices according as you ordered me & sent you a coppy on the other side. I straitly charged them to use every endeavour in their power to make it as good as possible as only such will do—But as I only began the advance from Tuesday, can scarce tell it at all, but in a few days we may judge the number of weavers already taken work out—and more quite ready if only warps can be sent for them they are all very good weavers.

"Mr. Yates will write to you to morrow by return of John but there is no doubt of him choosing to continue here, but as to him being usefull its wel known how usefull he ever was in assisting anybody but if he is to continue here I should be glad when you write to him to give him a caution not to asume to much authority over the work-people nor treat them with bad language not letting him know I have mentioned it and I'll do my best with him. Is it time to remit us in cash & bills about £60.

"Your humble Servt.

"THOS. HORSFIELD."

"ANDERTON, *December 15th*, 1788.

"SIR

"I am exceeding glad you sent me the Cash and Bills to-day as you will see by the Cash Acct. I should have been in great want to morrow & I think it will all work to get Saturday over with this. I have much wanted the list of prices for Winding you said you would send me, the Winders here will not work at the prices I offer them which is at 50, 5/7½ for 10 lb —60, 7/9—70, 10/7—80, 14/3 rising gradually in the intermediate numbers—but have told them I expected a list from you every day which we should pay by—so I should be glad if you would send it on Thurs or first convenient opportunity & on this account have not brought the Winding into the account. We keep warping what bobs comes in though at presant we have a bunch or two wound beforehand for I have not yet heard of a boy that is likely to do any good in assisting me—therefore am oblidged to take Wm. L of to day though we have many warps wanted for Wednesday—but he complains much of his leg being very painfull & says he is afraid he shall not long have it in his power to do any Service here. A person named Robt Crompton hath applied several times to come to warp or assist. He would either attend 3 day or the whole week at 21d per day Winter & 2/– Summer. He is as far as I can learn of an Honest Creditable Family & as warped for T. Swift but is now waiting till I have your answer. This morning Mr. Threlfall calld & desired me to let you know he would serve you weekly with what cash you would take at your own price but I can pay the spinners bills about £20 and allow them 4/– with wh they will be well satisfied—so that if we had cash to pay the weavers we could do very well I think about £50 a week. This week Swift has took 5 of our weavers. The reason is they are unwilling to knit their own healds—most of them have got liberty to weave a warp or 2 in his geers & now he tells them they must bring them in or fetch work from him, some of them refuses to pay for their worstead & other pays very reluctantly. We are badly of warps. The weavers are pushing against Christmas & much afraid of waiting.

"I am Sir

"Mt Ob St

"THOS. HORSFIELD."

"We want more paper to pack with.

"Mr Sam Oldknow Stockport."

Chapter IV.

THE LONDON WAREHOUSEMAN

THE letters of "our Mr. S. Salte" which furnish so much of the material of this book have a three-fold interest. The writer is revealed not only as an individual with a vigorous, lively, and not unengaging personality, but also as a well-recognizable type of business man and as the representative of a special system of commercial agency. The first of these aspects it is not very easy to disentangle from the second. S. Salte was clearly a born hustler, and the energy, shrewdness, and self-confidence which are fundamental characteristics of that class often carry with them a breezy optimism, a genial frankness, and a jovial humour which are not inconsistent with a certain opacity of mental and moral epidermis, and are scarcely more indicative of individual character than the good bedside manner of a doctor. But to us at least the heartiness of S. Salte seems more genuine than this, his bravura more unaffected, his moral reflections, though tending to banality, more sincere, and we may surely claim his familiarity with Milton, Gray, and Cowley and his use of them in commercial correspondence as an amiable idiosyncrasy of his own.

These mannerisms, however, whether personal or professional, call for no editorial comment. They display themselves to the reader and will, it is to be hoped, relieve the tedium of business detail. It is the third aspect of Salte's correspondence that most needs an introductory word. To describe S. & W. Salte as a firm of London middlemen does not carry us very far. Nothing was more characteristic of 18th century enterprize than the growing variety of its middlemen. An American scholar has recently attempted to distinguish between these varieties—the factors, jobbers, salemen, merchants, etc., who offered their services in the corn, coal, cattle, and textile trades to link up a growing range of producers with a widening market of consumers. London was then the main focus of their operations, even in the textile trades, and a country manufacturer had more chance of finding there than anywhere else a middleman suited to his special circumstances. In prosperous times indeed the appropriate agency might seek him out and offer to support him with its credit. This as we have seen is what had happened in the case of Oldknow.

But what was the special character of the middleman function undertaken by S. & W. Salte, and what was the nature and the extent of the support which they furnished to his enterprize ? To a large extent the answers to these questions will be found in the letters. In them and in Oldknow's day-books S. & W. Salte appear as a firm of warehousing merchants whose main business was to buy from country manufacturers and from importers and to sell to retailers.[1] They had to keep in constant and sensitive touch with the fluctuating demand of fashionable consumers on the one hand and with the technical resources and business capacity of an ever-changing body of manufacturers on the other. They had to be prepared to offer credit both to manufacturers and retailers, and the extent to which this could be extended in any one case constituted perhaps the main problem of the business. The great "Linen Houses," to which class Saltes' probably belonged, are said to have needed a capital of

[1] The best general account of this important aspect of commercial history during the previous century is given in Dr. R. B. Westerfield's *Middlemen in English Business. The Letter Books of Joseph Holroyd and Sam Hill* 1706 *and* 1738, edited by Prof. H. Heaton for the *Bankfield Museum Notes*, Second Series, No. 3, illustrate the operation of the middleman function in the Yorkshire woollen industry.

£30,000 to £50,000 for their establishment. This would have sufficed to set up a first-rate captain of industry. It was mainly invested in goods bought and awaiting sale in their warehouses or in debts owed by retailers. In July 1786, for instance, S. & W. Salte received over £3,000 worth of goods from Oldknow and had large amounts invested in other English and Scotch manufactures, in Irish linens, and in East India calicoes and muslins. But Salte also speaks during the crisis of 1788 of heavy debts owed to them by retailers. Other linen houses with less discretion had embarked larger sums in this form of enterprize. "The superb Retail Shops," says a contemporary newspaper, "which have arisen in every corner of the Metropolis have been chiefly supported by the Wholesale Houses without any capital of their own," and it speaks of the "Expenses of these Shops, the number of their Servants, the high Rents, enormous Taxes, and by the Rivalship the low profits," as contributing largely to the crisis of 1788.[1]

S. & W. Salte's business relations with Oldknow, though nowhere precisely defined, are rendered fairly clear by the letters. Salte is not a factor or commission agent. He is a merchant who buys the goods outright on certain conditions. Some are definitely ordered and cannot be returned unless they are defective. In addition to this Oldknow is authorized to send an indefinite quantity of specified kinds of goods, and S. & W. Salte not only undertake to accept these but are angry if they are sent elsewhere. But the kinds so specified are partly a matter of somewhat vague understanding, and this is liable to variable interpretation in different conditions of the market. Against certain kinds of goods Salte is always protesting, *i.e.* calicoes and prints. Others are included in this category as competition gets keener, until in the

[1] *Manchester Mercury*, May 6, 1788.

end the exceptions become the rule and Oldknow is told to stop his looms unless his weavers can make better goods.

Such remarks as these are due to temporary impatience and may perhaps be regarded as "hot air." Apart from an admitted but unexplained exception in favour of Parker, Topham and Sowden, S. & W. Salte claimed to be Oldknow's sole agents in London and any London firms who applied to the manufacturer direct were to be referred to them, though they might be rejected as customers on the ground of their insufficient credit.

S. & W. Salte to S. Oldknow :

"London, *February* 11, 1785.

"We have to acknowledge two Letters from you. The engagements at the India House, & other matters, have prevented an answer sooner from S. S. We left Bath the Saturday after we parted at Bristol not a little tired of the buisiest & Idlest place in the world.

"As for myself, the same faces, and the same amusements grew very tiresome, and yet I do not repent the Excursion, as the Contrast makes real business & retirement more pleasing—but Nancy was full as willing to quit this place of amusement, where fancy is perpetually on the Wing. We reached London on Sunday Evening all well. We thank you for your acct of your Journey, it was more expeditious than we apprehended.—The Historian of Birmingham I find was a proper Gentleman to address you to—You was highly favored with dancing with his Daughter at the Ball.—She I beleive is a Sensible young Woman. the Graces do not play in her face. She would be proud of her Partner, & Soft things must pass upon the occasion, but I ask not the result. Now Sir, to matters of another kind. We shall continue to send the money every Saturday as usual—if our Bankers shd complain of the Trouble will let you know. I was fearfull they would when I talked with you at Bath upon this Subject—We wish to render you every Service, the Principle we set out upon, and we think yours a reciprocal one.—You have had Mr Wadham with you. he failed abt 2 years since, & we refuse him Credit. The other, Thwaites keeps a ticketing Shop. These last offers are your worst—We confide in your honour to send us all your Muslins & as fast as possible, & we wish to give you every hint in the prosecution of your Manufacture. 1500 Lots in the private Sale that begins next week. Many Balasore hkfs thick & thin. Neckcloths a great many, Demy 6/4 Striped & Checked & yard wide demys a large quantity. demy 5/4 & 6/4 Jaccts a very large asortment. This does not frighten us. Go on & improve your Manufacture. you have the first Name & you shall not lose if our Industry & fortune will weigh any thing in the Ballance."

S. & W. Salte to S. Oldknow :

"London, *April* 23, 1785.

"Your new way of Numbering with small Letters will not do by any means. The old way is much the best, A B etc. This will puzzle our People beyond measure. The Aprons & hkfs are not got up in your usual way, they have the appearance of Scotch. You do not like it yourself. Keep to the old method if you cannot improve it. Where are the Romall hkfs ? Where are the 6/4 & 7/4 Stripes & Checks ? We are sorry the Spring Trade is Slipping away very fast, & not a piece of these goods, if you sell them in the Country, or Supply other Persons with them, it will totally disappoint us, & prevent us being the Friend we have been. Surely you will not do any thing ungenerous. . . . We congratulate you upon the Event of the duty being taken of[f] your Manufacture, it is glorious, & we hope you will improve the Golden Opportunity that now presents itself, of improving every Article to the Uttmost."

S. & W. Salte to S. Oldknow:

"London,
Saturday night, *March* 4, 1786.

"The Goods which were to have arrived yesterday morning by Pickfords waggon are not yet arrived, a sad disappointment to us. We have sent to the Inn but can procure no certain Intelligence—This is wrong. It is an injury to you & ourselves. You must try to redress this grievance as early as possible for much depends upon punctuality—£200 in Cash was sent last tuesday morning & hope will come Safe. Now to business, the Napkins are too short & narrow must be made larger each way, & to imitate the India a Red Stripe or border at each end, observe this in future. . . .

"The 9/8 Calicos for printing are sold so low that we can do nothing with yours. As good are sold at 21d p yard 4 [months] Credit & 5 P Cent, & the printing Trade is in such a wretched situation at present, that these all lye upon our hands. Turn your Weavers to Muslins. [To] forward this Manufacture we now enclose you some patterns drawn [from] Different articles of Muslins, come over as presents to the People of Fashion. We also enclose you two patterns of Muslins, the patterns are written under to describe to you how they are Worked. . . . Yesterday S. Salte waited upon Mr. Eden the Comercial ambassador to France by his desire, intreating him to furnish him with an acct of the situation of particular French Articles. S. Salte mentioned to Mr. Eden the infant Manufactory of English Muslins among other things. Mr. Eden was so Pleased with the acct that he begged S. Salte to attend the Lords of Trade next Wednesday at Whitehall to give his Evidence upon this & other matters. S. Salte has received a Letter from their Lordships in consequence & means to attend them & exhibit to them Samples of Mr. Oldknow's ingenuity.—Will not this animate your Zeal to excell? In a future Letter shall let you know particulars of this Examination before the Lords of Trade—We do every thing for you, you must do all you can for us. Send all the Muslins to us."

S. & W. Salte to S. Oldknow:

"LONDON, *March* 15, 1786.

"We recd your Letter this day inclosing a Bill value £171–8 which have placed to your credit—yesterday we sent you by Pickfords waggon £200 in Cash—We hope it comes regular. We are glad you have given Strict directions abt the delivery of your Goods in time, as we have Suffered many disappointments.—You are the best judge what patterns are fit for the Loom. Do not attempt any that you cannot excell in—as Men we are taught to imitate the Virtues of Angels, but we fall very short. Yet the brighter & more perfect the Exemplar, the more ardour & ambition we should indulge in our low & imperfect situation—Genius receives great help from Ambition, & despair is not our Motto, or any part of it, nor must it be yours. If we can send you from time to time new hints to improve your Manufacture, we shall do it without solicitation. We consider ourselves as embarked with you upon the same bottom, & we must sink or swim together . . . S. Salte has had a long interview with the Lords of Trade. They hardly knew of a Manufacture of Brittish Muslins—I took with me various Samples. I need not tell you the Complts that were paid They were equally surprised & pleased at my Exhibition. I was under Examination upon this Subject full two hours—I am promised a Visit from some of their Lordships. Mr. Eden was present all the time—I have copyed for your perusal on the other side, a Sort of Memorial I presented to their Lordships to read over previous to my examination containing a brief history of the rise & progress of the Callico & Muslin Trade in this Country—you will not give a Copy of it, as their Lordships have it, & it might not seem well to get into print any way, it was done in great haste & with imperfect knowlidge—I am in a few days to be examined by their Lordships upon the present state of the English Linen Manufacture. I have written for Documents to various parts. Mr. Eden will depart for France, so soon as the needfull information is obtained upon various matters of manufacture & Commerce. Let us hear from you again soon.

Observations on the Brittish Muslin & Callico Manufacture read & delivered to the Lords of the Council for Trade.—

"These articles were unknown in this Country a very few years since. but the progress towards perfection hath been very rapid indeed. The mechanichal Skill of Mr. Arkwright & other persons connected with him, in erecting Mills & Machines for Carding & Spinning Cotton Wool is the primary cause of all improvements in the Cotton Manufacture. The first Efforts of ingenuity in this new art, was directed to the manufacture of low priced Callicos for the printing branch of Trade an[d] other articles that did not require an attenuated Thread. Succeeding in these Fabricks, their Ambition & ingenuity pointed at something still higher & more important. The Object they grasped at was great indeed—to establish a Manufacture in Brittain that should rival in some measure the Fabrics of Bengall. This was treated by many persons who had great knowledge in Bengall piece Goods as a very wild & Chimerichal Scheme—but it hath not turned out so. Three years experience hath more than justified their most sanguine hopes of Success in the Callico & Muslin articles. The Scotch began first—they took the lead in this infant Manufacture—& though every degree of patient Industry must be allowed them, they have not been equally Successful with the Lancashire Manufacturer. With the patronage and encouragement of the State, it is not possible to ascertain to what an extent in quantity, or to what perfection in Quality this Infant Manufacture may arrive in a few years. One obvious Inference may strike your Lordships upon this Subject—that the rise of the Cotton Trade hath caused the decline of the Silk Manufacture in this Country, & if the French adopt the Cotton trade (as no doubt they will) it will operate with equal power against the Silk Manufacture of France as it hath done in Brittain—

"I hope your Lordships will pardon this imperfect Sketch I have here given you of a new & almost unprotected Manufacture—I feel my own weakness as an Advocate upon this Subject. Your Lordships have a compleat one in your own bosoms. If your Lordships consider this Object in a limited & local view only, as giving Bread to a part of the Laborious poor of our own Country—it rises very high—but if your Lordships consider it in its remote & future consequences as enriching the Manufacturer & Merchant, & yielding an ample Revenue to the State—it rises infinitely in dignity & Importance."

S. & W. Salte to S. Oldknow:

"London, *April* 4, 1786.

"We received the Small Bale by the Coach in which were Bills value together £1095–1/– which have placed to your Credit, we hope you have received the £200 in Cash regularly we sent the same Sum this morning by Pickfords waggon—We wish you to quit the low

Ballasore hkfs, they will not answer. We have now as we think made a fair Experiment, & they will not answer. The Scotch bear the Bill in this article, & you & we must be content, if we rise Superior in fine Goods. press forward in all the finer Articles & as fast as possible—We wish you to Sell the Callicos in the Country. Strutt sells his at 20½ [which is] Free of Carriage, another Gentn at Stockport offered us this [and] any Quantity of fine White Callicos at 20d free of Carriage—You see how this article is situated & we presume you will not pursue it at present, indeed we purchase none at these low prices—for we do not know what sort of work we can turn them to that will be profitable to us.—Coward of Bath is in town—he wonders you have not availed yourself of his patterns & made some Goods from them. We told him you could not conquer Winter weather. We want as many Spotted Muslins & fancy Muslins as you can make the finer the better.—You have many Competitors we hear coming forward, indeed Aprons have appeared well made & very nice. You must give a look to Invention, Industry you have in abundance. Send by the Coach every day what you can. this Month & the next are of the uttmost consequence to you & to us. We expect to hear from you as often as possible & as the Sun Shines let us make the Hay."

S. & W. Salte to S. Oldknow :

"LONDON, *April* 19, 1786.

"Yesterday we sent you £200 by Pickford We recd your Invoice by post this day. The aprons came yesterday. only are a Drop in the Buckett. the low priced goods make but few. Some Complaints we have had, & many Spies. Parker Topham & Moore Sell all your Goods cheaper than we do, or Else folks tell lyes. We take up with little profit, and some not well made a very trifle.

We wish no person to interfere with us, it is a great injury to us & to you for any person to depreciate & sell very cheap your goods. S. Salte was with the Lords of the Council again yesterday, but not time now to relate particulars. Some talk about you & Arkwright—let us hear again soon."

S. & W. Salte to S. Oldknow :

"LONDON, *May* 10*th*, 1786.

"We received your letter with the Bills very Safe, & have credited your account with the same—We sent you £200 by Pickford this day, which hope will come safe—the Muslins & Sattinetts came by the Coach this morning. We rather wish you to drop the Sattinetts, they are not new here & only fit for 2 months Sale. The Buff Stripes are liked best but still do not pursue it, turn the Loom to something Else. They are not fine enough for people of Fashion, for which they are only calculated for.—We want more Napkins & some made larger—try your skill at Table Linen . . . Arkwright must lower his Twist & he must Spin finer, tell him the reputation of our Country against Scotland is at Stake—We hope to see you in town soon & shall collect all the new hints in our power to improve the Manufacture of Muslins. Young Strutt of Derby is in Town, they are sadly sick of the Callico trade & the other articles they have made. Great Revolutions we think will happen in Lancashire amongst the Manufacturers of Cotton yarn."

S. & W. Salte to S. Oldknow :

"LONDON, *May* 23, 1786.

"Our next India Muslin Sale is fixed for the 29th June, if you wish to have a peep at the Goods you must come to town a week before. Our or your Goods have still gone of[f] to [ ] well in spite of all opposition. You have Enemies & we have Enemies in abundance—Some to dispericate & others to rival us, in better goods they try at a lower price. Our Ears are stressed every day with the Excellence of Scotch & Lancashire Muslins, if cheapness proves any excellence they have it indeed. We listen to the repeated Tales with a degree of Phylosophy that would not displease you if you heard them, for a mixture of Praise now & then comes in by chance. The Scotch Impudence & perseverance is beyond all, but there is a Manufacturer in Lancashire makes Excellent goods in Muslins & sells them by Comission in London, we cannot get at his name. A Friend of ours last Saturday bought £300, all that came. amongst them 9/8 & 5/4 Balasore handkerchiefs of a nice thin & curious Texture, also Aprons, he says Superior to yours & much lower. Perhaps you may learn from what quarter they come. . . . Try what you can do, we well know you are surrounded with Spies. Every man [tries] to Steal a little from you. We are also not without our Enemies who circumvent us in our [Sales] & try to recomend other Makers. Let Ambition fire thy Mind to Excell every man, every hint & every assistance you are sure to receive from us, & pray let us have the Goods, & send none of the Patterns we furnish you with hints for to any other Person. It hurts us, it Degrades the sale."

S. & W. Salte to S. Oldknow :

" LONDON, *June* 5, 1786.

" We have no Invoice this day, the Goods by the Coach came on Saturday, and the Goods by the Waggon only came on Saturday morning. This inconvenience should be remedied & deliver'd to us on Fryday morning a much more leisure day—The Sattin Stripes in colours will not do, & we desired you not to go on with them sometime

Since. We allways expressed a dubious opinion about them.

"The Scotch have done much better this two months in the same. Do not make any Shawls, they are totally unsaleable, we mentioned that some time since. We are dying some of different colors to try what that experiment may accomplish in the Sale of them. The Striped muslin Cravats are too course & thick, drop them entirely. We hinted at first our doubt about any demand—low 6/4 Striped will not do any longer, the Scotch have routed them out. Fancy broad Stripes are the fashion & good Quality from 42/– to 60/– . . . indeed the Scotch perseverance & ingenuity are doing wonders. the Lancashire People are all exerting themselves, & no less than three or four houses opened to sell Lancashire Muslins by comission, & the accounts of buyers are much in their praise for Execution & Cheapness. The Scotch have sent up many Spotted Muslins, indeed too good & too cheap – You must make them thinner & raise the Spot. different sizes of Spots, Barley corns, little Clubs, and any other [ ] Figure will vary & give more choice—they answer for Cloaks as well as aprons & gowns.

"We are buying fine & excellent 9/8 Callicos at 20d & 4/4 wide at 12½, 4 months & 5 P Cent. What are we [to] do with yours. What can be the reason of this? We are willing to take 600 pieces of your Number 4 Callicos made as usual at 20d & then we have Carriage to pay. The Printing Trade is so over done here, that we cannot take more at present & we wish to encourage no other Maker. You have handkerchiefs, Callicos enough. Stop your hand at Present & from time to time we will give you fresh information. We expect you in Town about the 20th or you will be too late to see the goods at India house. We expect the Jaccts fine hkfs, fine checks & Spots by Coach every day, as fast as can."

S. & W. Salte to S. Oldknow :

"LONDON, *June 13th*, 1786.

"This day we sent you £200 by Pickford & hope will come safe. The parcel by Coach came this morning. Private Trade India ones are now selling very cheap, & nice goods of the Decca kind—Many wholesale drapers now give out they have goods coming up equal to yours which they are to sell by comission—We trust our *Laurel Tree* will not yet fade. We dare your rivals to the Field. If they should bear away the prize, they must fight hard and deserve well indeed. Mr. Strutt Senr was here this day. Have had much Conversation with him & part relative to you. . . . "

Sam Salte to S. Oldknow :

"LONDON, *July 12th*, 1786.

"I have this minute closed my Book for the Muslin Sale, & am going to ride down to Rotting dean to see Nancy who has been there near 14 days. But to business as I am in haste, all fine thin Goods have sold dear enough, low & mid price thick Goods cheap enough, Striped Dorcas buy cheap indeed, the Check goods higher—but as there was no 6/4 wide ones it is an advantage in the Sale of yours—the Single two & 3 Cord Checks we told you are more Saleable by far than any others & that must be attended to most. the Scotch have loaded the Town so much with Ballasore handkfs that desire you would not send any more at present except fine 9/8 & 5/4 wide handkerchiefs & the finer the better, & also fine 4/4 & 9/8 Checked hkfs. . . . I would not advise you take Arkwright's Callicos. What will you do with them in these times ? You need not send any Goods by the Coach, as the waggon will bring them fast enough.—"

Chapter V.

THE YARN PROBLEM: NEGOTIATIONS WITH ARKWRIGHT

IN passing to an account of Oldknow's relations with the spinners a few words on technical matters are unavoidable. The coarser and finer varieties of cotton yarn are distinguished as low and high counts respectively, the number of the counts representing the number of hanks of yarn to the pound weight. For the weaving of muslins comparatively high counts both of warp and of weft are required. The great achievement of Arkwright's water-frame was that by the twist it gave to the thread it made it hard enough for warp, and warp was accordingly known as twist, or hard yarn. The jenny could only spin weft, but it could produce much higher counts of weft than the water-frame could of warp. The price of any count of warp was higher than that of the same count of weft, and as the count became higher the difference in price became greater.

In giving an account of his early life to Mr. Bannatyn, Crompton stated that "soon after the invention of his machine he spun a small quantity of No. 80 to show that it was not impossible . . . and that for the spinning and preparation of this he got 42/– a lb."[1] Now, in Oldknow's stock-taking of 1783 there is mention of 6lbs. 57 hanks of No. 84 twist valued at 44/– per lb., and of 21lbs. 11ozs. of No. 94 twist valued at 50/– per lb., whilst the highest count of weft mentioned is No. 80 valued at 22/–. As the records have not revealed at this early period any other cases of twist so fine as eighties and nineties, it seems not improbable that the small quantities in Oldknow's stock had been spun by Crompton himself. They were almost certainly spun on the mule, but they can only

[1] Baines: *History of Cotton Manufacture*, p. 200.

have been used for experimental purposes. In his calicoes, shirtings, and sheetings Oldknow did not use, in 1784, any higher counts than No. 44 in twist and No. 47 in weft; and even for muslins he did not at first get beyond No. 66 in twist and No. 86 in weft. We have what is apparently a complete account of the twist delivered at Anderton in November 1784. Out of a total of 1,620lbs. only 20lbs. was in counts over 60 (the highest being 66), and three-quarters of it was in counts of 28 and under. There can be no doubt that the greater part of this was produced on Arkwright's machines. Fine counts of weft as high as 70 were being supplied to Oldknow in 1783 by spinners of Bolton and Ashton and by merchants in Manchester who combined dealings in cotton and yarn. One of the advantages of Oldknow's establishment at Stockport was that it enabled him to deal directly with a large number of small spinners. The accounts of some fifty of these in the tattered fragments of a ledger of 1786–7 reveal a considerable variety of types, from the woman who earned a pound a fortnight by spinning low counts on a jenny to the owner of a small factory of mules or jennies who supplied fifty pounds' worth of weft in a month. Perhaps the broadest distinction is between those who were paid in cash, fortnightly or oftener, and those who were paid usually by bill every month or two months. All the former class and most of the latter received and were held accountable for the cotton but did not purchase it. Only in a few exceptional cases was the cotton sold to and the yarn bought from the spinner, and in one of these interest was charged, showing that the sale of cotton had been on credit. More than a dozen clearly specialized in high counts from 70 upwards to 92, and these lived chiefly in the Mellor district and at Ashton, Hyde, and Stalybridge.

But it was weft that all these small spinners were engaged

in producing, and whilst still finer counts of weft were required for the improvement of his muslins the chief technical problem that confronted Oldknow was to get finer twist of regular quality in large quantities and at lower prices. This problem was solved by the application of power to the mule, which by the end of 1791 was producing in Oldknow's own factory at Stockport No. 120 weft at 22/– per lb. and No. 120 twist at 23/–. At that date, however, Oldknow's career as muslin manufacturer was drawing to a close ; it was in the previous five years that he acquired his reputation and his fortune, and the beginnings of fine spinning during this period have hitherto been obscure. William Kelly, the manager of the New Lanark Mills, claimed to have been the first to apply water power to the mule in 1790,[1] and the establishment of Drinkwater's factory in Manchester, of which Robert Owen was the manager, belongs to the same year.[2] But the mule was in general use before power was applied to it, and William Radcliffe tells us that " the mule twist came into vogue for warp as well as weft " in the year 1788, and that this led to an immense expansion of weaving in the Stockport district. Our new records, whilst confirming this statement in the main, enable us to trace the beginnings of the new development.

As Arkwright's patent for roller-spinning might claim to cover the similar element in the mule, it is probable that the latter was not freely used till the final rejection of Arkwright's claims in the summer of 1785. In April 1786 Salte writes to Oldknow: " We hear Arkwright has lowered his twist," and again on May 10th: "Arkwright must lower his twist and he must spin finer. Tell him the reputation of our country against Scotland is at stake . . . Great Revolutions we think will happen in

[1] Baines : *History of Cotton Manufacture*, p. 205.

[2] Owen : *Autobiography*, p. 6.

Lancashire amongst the Manufacturers of Cotton Yarn." The clue to the significance of these hints is given in another of our records. The ledger containing the accounts of the small spinners covers the period between the middle of 1786 and the middle of 1788. In 1786 the prices show that nothing is being produced but weft. At the end of February 1787 some of the yarn is distinguished as "hard" or "twist" and for this higher prices are paid. The counts of twist thus supplied are mainly those between No. 40 and No. 60, which were amongst the highest hitherto produced by the water frame. It is clear that the reduction actually recorded in Arkwright's prices in June 1786 was due to the competition of the mules in the hands of small spinners, and that the changes whose fuller operation is ascribed by Radcliffe to the year 1788 had their beginnings in the summer of 1786.

The emergence of a new and freer source of supply of materials for fine cotton fabrics affected the established interests not only of spinners but of manufacturers, and tended to draw both together for mutual defence. The overtures which we shall find being made by the Peels to Oldknow were a gratifying recognition of the rank he had attained as a manufacturer, but he was already under obligations to Richard Arkwright, so that an alliance with him and his son Richard, who had a separate spinning mill at Bakewell, was the more natural solution of the problems confronting him. He needed the support of their capital and a steadier supply of twist than was obtainable in quantity elsewhere. The Arkwrights desired to retain the market for their yarn and to maintain and extend if possible the use of their machines in face of the new inventions. Of the negotiations that took place at Cromford, Chapel-en-le-Frith, and Bakewell during the autumn of 1786 the letters that follow give us the only available account. They do not tell us everything, but Salte's lively comments as a looker-on are almost as good as more exact knowledge. It is clear that the proposals under consideration included a partnership between Oldknow and the younger Arkwright and the building of a new spinning mill and that both of these plans were dropped. William Strutt, it may be added, was not a party to the negotiations but was probably cognizant of them. He was the son of Jedediah Strutt the former partner of the elder Arkwright, and later won some distinction as an inventor and as a man of science, being elected a Fellow of the Royal Society.

Sam. Salte to S. Oldknow :

"LONDON, *September* 18, 1786.

"We parted last at Cromford, after I think a very agreeable visit to Mr. Arkwright I suppose he would keep you longer as you mention & I believe he wishes to shew you every kind of Civility. A Civility that in its consequences will be productive of much Service to your Manufacture. Young Mr. Strutt told me in going to Derby, no other person could accomplish fine Spinning & such as you want ; the old Gentleman was sanguine or apparently so, that he would accomplish it—indeed my good Friend something must be done, to perfect this yet infant business—We find ourselves pressed very hard upon from every Quarter.—

"We want a Glaze or Dress upon the goods not yet accomplished, & upon every sort. There is a new invented Cylinder made of paper that does wonderfull well & gives muslins the India Gloss. Think, & think again on this matter. Indeed my dear Friend we have this Manufactory so much at heart that we must not be out done by any people, whatever. We Must not be Second to any house in Rome—First or Nothing. Let ambition have her full

Scope. . . . We hope times will mend, or the market less glutted with British muslins. *Fair & Softly goes farthest.* We know you will thank us for being explicit & candid, & do send us up the account of your Manufactured goods as you promised. And now my Dear Friend I must particularly request you to make my most respectfull Compliments to your Brother & Sister, & Miss Hall thank them ten Thousand times for all their kind Civilities to me at Heaton. I do asure you thepleasing hours I spent there hath been the occasion of many pleasing reflections. I only wish to see them & live those happy hours over & over again. Whenever I speak of an agreeable retreat, I always mention Heaton, Heaton. Is your house now finished? Have you done with Carpenters, Painters, etc. You will be glad to turn them out, & take in Somebody, sweet Somebody—The sooner the better, it is not good to live alone. My complts to the Revd. Mr Wilson."

Sam. Salte to S. Oldknow:

"London, *October* 11*th*, 1786.

"How did you weather the stormy night you left us? I think I never remember one more tempestuous. What a contrast a comfortable bed & a sweet complying good natured wife would make to it. The heart that knows the difference, would feel more sensibly your situation, get these things altered. Happyness cannot begin too soon, nor depart too late. This is a text to be enlarged upon at some future period, but to other business. We want some 10 to 20 dozen of huccabk napkins by the Coach for a shipping order etc . . .

"Pray my respectful complts to our Friends at Heaton. Tell them I often think how happy they are, but I dare not envy them."

R. Arkwright, Junr. to S. Oldknow :

"BAKEWELL, *October* 13, 1786.

"I am this morning favoured with your letter of the 10th. I wrote you by the waggon on Tuesday in answer to your two last letters which you would of course receive, and I may perhaps hear from you to-night.

"I do not know of any business likely to call me over to Manchester at present.—Therefore I send this to desire you will meet me at Chappel on Sunday morning, by 9 or half past, to breakfast, I know will not be too soon for you, and I will endeavour to be there by that time.—I have wrote to Mr. Simpson to meet me there also at 12 or 1 o'clock to dinner so that our different businesses will not interfere with each other.

"I observe what you say about this great business likely to be carried on in opposition, you may certainly call it, for it will interfere in some degree with you. It must be a long time before any new house can do the business you now do, with any comparative degree of advantage.—You will please to say by return of the bearer if I may expect you at Chappel at the time—half past 9."

S. & W. Salte to S. Oldknow :

"LONDON, *October* 23, 1786.

"You solicit my advice upon a very nice & difficult subject. With respect to the proposals that the Two houses of the Peels made you, it required no hesitation on your Part. I considered them as selfish in the Extreme, without the weight of a Feather to preponderate in your scale. Mr. Arkwright's proposals come in a less questionable Shape, & his character determines much in his favour. If I could turn only a few leaves of the dark pages of futurity, and take a peep at the remote consequences which may arise from this conexion with Mr. Arkwright my doubts would be removed, & a decisive answer ready.

In new schemes & projects the probability, if not the moral certainty of success must allways be taken into the account. A wise man would not venture himself upon ground that was dangerous, & likely to sink under him. Nor, as Sancho Panza says, would he go a Fishing for Mushrooms in the Sea. There are many previous considerations which your Judgment will weigh before you jump to the conclusion. Are there not more Cotton Mills allready than can find employment? What will be the probable amount of a new Mill, complete & ready to work? Will it not involve you in expense without paying you such Interest as you have some right to expect? Will it not interfere with your personall attention to a manufacture that requires all your time & all your Skill? These & a variety of other considerations will suggest themselves to your mind before you enter into the Engagement you speak of. If Mr A Junr would give you a share in the Bakewell mill, & if you was to give him a share of your Manufacture, might it not be for the mutual advantage of each? I think no risque could accrue from a Union of this kind, he would strain every nerve to improve the Spinning, & you would be equally alert in the improvement of your Manufacture —The Honour & the profit to be a joint & equal concern. I do asure you many persons here are going to sell Brittish Muslins by comission. Mr. Husey is going to sell for Ainsworth, & has allready exhibited very good things in handkerchiefs, aprons etc., some persons say more perfect, & at a lower price than ours. We perceive we shall soon have many competitors to contend with, & it will require all your skill & Industry to manufacture better than others, & our diligence Zeal & Eloquence to sell them.—We shall not easily submit to come in Second in this race. You mention drawing for £2000 before this month is out. We have your letter this instant in which you have drawn for £1700, we suppose that is part. May it not be prudent to stop many things which we mentioned to you, let us not repent too late, better keep the market bare than clogg it too much. £1200 a week at this season or before February we think will be too much, especially as the prospect is. Moderate a little—gently along. All your yard wide at least 20 pieces of Callico shirtings send up by the first Waggon, we have some chance of selling them to be transported to Botany Bay. This letter I hope you will receive before your interview with Mr. Arkwright. If upon an impartial review of my letter you think any observations not weighty enough to determine your opinion finally as to the Erecting a Cotton Mill you will let me know. My present Sentiments are against it. Bold speculations sometimes are fortunate, one successfull adventurer hath led hundreds to ruin—your Prospects allready are such, that nothing should be left to accident or chance—Ambition is a very active & usefull quality of the mind, when it is under the direction of experience & Reason. Why does not the Tide flow faster & higher, said a passenger in a Ship to a Sailor It would drown the Country if it did, said the Sailor with a Smile at his ignorance—You ask where you can borrow £5000—indeed I could recomend you to some persons who are now selling out of the Funds, but they expect Land Security & 5 P Cent. People in a large Trade have nothing to spare—After your interview with Mr A you will favor me with a letter containing particulars."

R. Arkwright, Junr. to S. Oldknow :

"BAKEWELL, *October* 31, 1786.

"I am sorry to trouble you about the linnen yarn which you was so good as to get your brother to bleach a little for dying yellow. I think you promised to get it me dyed at Stockport.—If it be done I must request you will let some of your men call for it and send it me in the wrappers this week if possible, for I believe they are quite without in the Mill.

"The quantity of twist sent you this week is rather under your order, but I think you said you had pretty well of all sorts, but about 50s and those numbers you will receive a good deal of.

"What day do you think you shall be here, if you come the next week? That I may not be from home.—The sooner the more agreeable to me, for I wish to come to a conclusion on the subject we talked of.—You can perhaps put an end to it by letter, by return of Bright, and then say if you please, whether 1 or 2 for 6 or 12 months certain will be of use, and it is at your service. If you do not come next week, and your determination makes it necessary, I will give you the meeting on Sunday at Chappel, but of this you will write.—

"I hope the twist is now very good, and I wish you would at your leisure some time soon, make out a fresh weekly order, conforming to your stock, for you must have a good deal of about 40s—I expected to have sent you some 14s of the new sort, and I believe it will be as good as the other, excepting having a little more dirt in it.—You will tell me what it is worth when you see it.—I believe we are to have an Assembly at Bakewell on tuesday, would you think it worth your while to be here on that day, if you could make it convenient?—"

Sam Salte to S. Oldknow:

"LONDON, *October* 31, 1786.

"I have been working at the India house all this day among the Muslins, & am now going out to dinner. Your letter does not require much consideration, as you & Mr A have brought matters to, or near it, an equitable conclusion. As you are both Men of honour, & understand one anothers tempers pretty well, you must not disagree about trifles in forming a Conexion, that I hope will add to your property & happyness. I cannot conceive of a more natural union than yours & Mr A's & I prophecy great things from it. I hold Mr A's character in high estimation & I do not undervalue the Merit of my Friend Mr. S. O. & I hope soon to congratulate both on an Event, that will have the best tendency to improve the Muslin Manufacture of this Country. You may defy united all competitors & all usurpers, but remember you must not Slacken your ardent Zeal in this race. You must both have the perseverance of Saints, & the resolution of Martyrs. Every possible hint & improvement that we can from time to time suggest to you, shall be done with great alacrity, as our Fate is in some respect involved with yours. Pray present my best Complts to both the Mr. Arkwrights, & tell them how much I approve of this new conexion.—You will let me hear from you after your Visit to Bakewell & Cromford that I may rejoyce in learning your terms of partnership are compleated.—We want some Particular . . . of Goods by the Waggon, & send us twice a week by the Waggon, but do deliver the Packs to Pickford that we have them regularly on tuesday & Fryday. Saturday is too hurring a day.—Give my best respects to your Brother & Sister."

Wm. Strutt, Junr. to S. Oldknow :

"DERBY, *November* 11*th*, 1786.

"I am much concerned that I have not been able to perform the promise I made you of calling upon you at Stockport the last month—but as my business in Lancashire would require a fortnight, the present extraordinary state of the Callico trade (on which subject daily applications have been made which required immediate answers) would not admit of so long absence from home—and fearing that from what passed at Cromford on the subject of my visit to you, you may have delayed the execution of some part of your plan, I just take the liberty to say that if you think I can afford any assistance or information & which cannot (perhaps from the nature of the circumstances) be communicated by writing—if the *time* is of the least consequence I will immediately go to Stockport any day you may appoint with great pleasure, and as it will require but one day I beg you will not hesitate to say what is agreeable—If it is any general information which you wish that may be communicated by writing I shall not lose a moment to give the best I can.—"

S. Oldknow to Wm. Strutt, Junr. :

"STOCKPORT [no date].

"I have been so long a letter in your Debt I must have thought it an honor. Your esteemed letter on the 11th of Novr deserves my best thanks for your kind offer of coming to Stockport to give us your opinion upon the subject [we talked] of at Cromford.—the Wheel is making and the plan of applying the water is nearly concluded upon—notwithstanding I shall be very happy to see you when ever it is convenient and shall take a great pleasure in showing you our Bleaching Ground and by making you acquainted with our experimental proceeding in that Branch—I do *think* you should have a bleach field of your own *you* would find it a great convenience, but come and see ours and judge for your self—Are you going to London soon ?—I purpose being there about the [torn] 15th inst—My present intention is to ride my horse to Derby or Nottingham (perhaps to both places) and then take the Mail—was you going I should be very glad of your Company to go either a post Chaise or the Mail, and I would endeavour to suit my time to yours. Sunday or Monday the 14 or 15 will be the time I think of leaving

Derby or Nottingham—I shall stay in London about six days, should my plan be such a one as suits you—I should be happy to receive a line from you that you will join me."

S. & W. Salte to S. Oldknow :

"London, *November* 18, 1786.

"We received your Invoice & letter this day enclosing two Bills value together £1520 to your Credit. We observe what you say respecting your treaty with Mr Arkwright, but it is not particular enough for us to Judge off. You certainly have not receeded from the first proposition, & indeed Mr A's offer to bring into your Trade £15,000 —you did not mention this in your letter—we apprehend it is to be so. However, when we receive the whole, & the apparent Necessity of your taking in a partner to conduct the Muslin manufacture upon a respectable & large Scale, we know of no person so Eligible in all respects as Mr A —— Junr., the result we hope will be happy & fortunate for you both. As far as it lies in our power to procure Success & advantage to you both in this new Conection, we will most readily comply, well knowing at the same time you will endeavour to make the concern of some consequence to us. Our Interests must be reciprocal. S Salte waits with impatience the conclusion of this very important Negotiation. Copaes sell at the India Sale much as the last, & also Callicos, & the very large Quantity must keep your Callicos down. We hope to hear from you again soon."

S. & W. Salte to S. Oldknow :

"London, *November* 21, 1786.

"We recd your last stating in part particulars of your agreement with Mr A, we cannot foresee any cause that you will have for repentance. We think there is a strong probability that this Event will be fortunate to both contracting parties. A mutual desire to please (as in a matrimonial conexion) must be cultivated, & harmony & Success will be the result.—When you have *done the deed* you will state further particulars."

Sam. Salte to S. Oldknow :

"LONDON, *November* 23, 1786.

"From whence, or what cause, arises this sudden revolution, in your Mind, respecting the treaty with Mr A—— Junr. I will not be inquisitive. You are a perfect judge of your own concerns, & certainly have weighed every circumstance in an exact ballance, *& better not do the deed, than repent it done*—Doubts I have entertained myself respecting the Event, whether it might be so fortunate as your imagination had painted it. My maxim was allways for a man to have the Control & management of his own Concerns, without admitting a Colleague unless upon a principle of Necessity. I think you was never reduced to this, & your Industry & ever Active Mind to make improvements in the Muslin Manufacture, will in a short time make you perfectly easy & independent. Why give the advantages of a Harvest to another, which your Labour & Skill hath been employed in raising.—Your request is granted with respect to drawing upon us. We feel a pleasure in supporting a worthy Man, & wish to lead him to the Temple of Fortune. You have grasped at too much buisness; reduce it, & you will be easy & happy. The only Chance you have of Excelling in Muslins is to contract the manufacture, & we look up to your Genius for Excellence in this trade. We will take Arkwrights' Callicos at the price you mention, & send them by the first waggon—We shall want your 3½ No Callicos when ready. Robt Peel says at the present price of Wool they only stand in 18½—the price is going down—I shall be anxious for the result as to Mr A—— Junr, & let me know as soon as you can."

Sam. Salte to S. Oldknow :

"LONDON, *December* 1, 1786.

"How true it is, I know not, but there is an old proverb, that Fortune favors the Bold—This I know, that she generally favors the prudent & the wise. I think in your declining the late Conexion, you will in your own mind be much happier, independence is seated there, & not in the regions of fancy & imagination. Those who make haste to be rich often stumble in the road, & Cowley the Poet, has an observation, 'that from a short life we should cut of[f] all hopes that grow too long'—but I cannot moralize any more, my attention being called of[f] to several objects & short in time. I apprehend Mr A—— to be what he has proved a Gentleman & I think he will assist you at any time in the pecuniary way. But pray above all do not involve yourself in too many Schemes, & too much buisness—it will defeat the very purpose you was born for, to live happy, & become independent. Drop the Chk Trade, & all the branches that a common Manufacturer can do better. Keep & confine yourself to the improvement of Muslins particularly. You will have rivals enough to contend with—& your Work must be superior. Watson who wants to hire your men makes admirable aprons allready & sells ten P Ct lower than we can. How this is we cannot explain, part with no good hands, let him have the refuse. You can employ & pay them as well as others. You want us to send £400 a week in Cash—We cannot do it on this side Xmas. our India Sale hath drained us at present, & a time of the year when we receive little, your Bills upon us will find Cash. I warrant them. We have sent by Pickfords waggon 4 Baggs of Cottn Wool, came from Barbados the weight

574lb charged us, give us what you can, the Bags are charged 7/6 a Bag 30/– & permit 3/10 let us know wt price we must charge your acct with. The Coach parcels, did not come regular, part not come in, nor by the Waggon. Invoices to come sooner."

R. Arkwright, Junr. to S. Oldknow :

"BAKEWELL, *December* 19, 1786.

"The Mill-wright has never called, tho I have been expecting him every day.—He will certainly be here in a day or two.

"You will receive this week a pretty large assortment of twist under 50s, and you shall have more 28s, and about 42s the next, and of a good sort too.—Can you make use of more 14s ? I have sent you two samples of the cotton they are made of. What can you afford to give for it, in either of the states you see it in ? Or what do you think it will fetch ? I have 40 cwt. of it, and I would be working it up for selling it, during the high price of cotton.

"You will favour me with a line by return of Brights man, and you will either remit me or not, as best suits yourself.

"It gives me great pleasure to find your warps are so well liked, and I am very glad you [ ] them so well. I have thought the twist good for a good while and it will continue so, or as you say get better and better. I am sure you will not make any into warps but what you are convinced is good.—You may depend upon having as much as you order from 38 to 50, if what is sent be not enough, therefore say what, and do not buy of Clough, or any body unless you can have it so much cheaper."

Chapter VI.

THE CRISIS OF 1787-8

OF the arrangement ultimately made between Oldknow and the Arkwrights no explicit account has been discovered, but its general purport is clear. The proposed partnership with Richard Arkwright, Junior was dropped, but Oldknow received large advances in the form of yarn and drafts which by the beginning of 1788 amounted to over £10,000. It must have been the confidence derived from this external support as well as the strength of his own purpose that induced him to take a decisive step towards the adoption of the factory system on a much larger scale by the first extensive purchase of land and water power in Mellor in the spring of 1787. His uncles and his grandfather, to whom he had no doubt owed much of his start in life, were now in need of his financial help, and the last of the five following letters renders it probable that Samuel's loans to his grandfather were not all repaid.

Joseph Oldknow to S. Oldknow :

"NOTTINGHAM, *September 7th*, 1786

"DEAR NEPHEW,

"Yours of the 4th I duly Received & shall be obliged by your sending me two Hundred Pounds of your Drafts as Under by Return of Post. I sent you yesterday, by a Parcel that went from Bro Tom's, Forty Pounds in Cash. You May all so Depend of the whole of the Cash in course of Next Month for the Drafts you now Send me I will in future take care to ask you no more for any Drafts, and if I can at any time be serviceable to you it will Give me Pleasure. I all so thank you for your Advice and will in future take care to Adhere to it. If you at any time want two or Three Hundred Pounds of Bills Discounting if you send them me they shall be Immediately Done. I understand my Father has wrote you by this Post. He is this Day to Meet Mr Abney about the sale of his Derbyshire Estate. If they Agree about it as I hope they will it will make my Father more Easy in his circumstances and bring his Affairs into a compass that will Ease him of a Deal of Anxiety of Mind ; if I Live and am well I will [visit] you some time Next Month."

Thos. Oldknow, Senior to S. Oldknow :

"CALVERTON, *November* 28, 1786.

"DEAR GRANDSON,

"Your kind favour received which gives me great pleasure and lays me under verry great Obligations for your kind assistance in raising me the Money at London. I doubt not but in a verry little time I shall be able to give you a more certain Account of my proceeding with my Friend and Neighbour Mr Bingham and am in no fear but I shall be able to furnish One Thousand pounds at May day next. I shall fully expect and be very glad to se[e] you as profered in Your letter but in all probability [I] may have occasion to write to you before that time and shall be glad of a line from you as often as convenient."

Joseph Oldknow to S. Oldknow :

"NOTTINGHAM, *March* 31*st*, 1787,
Saturday Evening, 6 *o'clock.*

"DEAR NEPHEW,

"I have just Received an Account from my Brother—from Calverton of the Departure of my Good Old Father he Resigned his Breath this Afternoon about Four o'clock without Sigh or Groan. As your Brother and you I Apprehend are Left Executors I could wish one or Both of you would Immediately Come Over. I have not time to say more being a Busy Day that we may be allways Ready is the Sincere wish of your Affectionate Uncle."

S. Oldknow to Thos. Oldknow :

"STOCKPORT [Undated].

"DEAR BROTHER,

"I should be glad [if] you would come & dine with me that we may consult on the above Letter."

S. Oldknow to Thos. Oldknow :

"NOTTINGHAM,
Sunday morning, *October* 28, 1787.

"DEAR BROTHER,

"Last night Mr Heywood my two Uncles and myself had a meeting the purport of which was to put my Grandfathers affairs into a way to be made an end of, and after we had got through that we tryed how much the Bond & Simple Contract Debts would receive in the pound. It appears there may be 10/– & more if Gransby Estate sells well—this matter and my own business made it impossible for me to have proceeded on my journey yesterday—and wishing to see Pleasley Mill which cannot be to day I purpose leaving this place early to morrow morning.—"

A still more serious cause for hesitation was the depression in trade which set in before midsummer 1787 and the threat of a crisis which was not completely realised till the following summer. The approach of one of the most disastrous epochs in the commercial history of Lancashire is reflected in the letters that follow. Salte had been prophesying trouble before Christmas 1786 and continued this Cassandra strain in every letter, whilst

Arkwright was not convinced till six months later. Oldknow's letter to Arkwright in June (one of the few that have survived) strikes a sombre but not a despairing note. The worst was still to come.

S. & W. Salte to S. Oldknow :

" LONDON, Undated [*Oct. or Nov.* 1786 ?]

" We have your favor from Cromford, the particulars of this meeting we expect in your next—we can say nothing new upon that Subject—we wish the completion (if one) union may eventually turn out to your Interest & happyness—as for ourselves we can have no wish but to add to both—we hope you complain without Cause.—Speculators in Cotton Wool & Mad Temporary Agents, we think will repent & very sincerely, there is no real cause for all this—not a single piece of Callico is really sold. tho' at the advanced price, they are laid up snug in Warehouses to come into the Market by & by—they will never suit the French Trade nor any other Trade at the Price, & must come down, & many *Persons down with them.* The next year will prove a very serious one to many imprudent persons, & perhaps the year after worse "

Sam Salte to S. Oldknow :

" LONDON, *December* 15, 1786.

" I received your very reflective, contemplative letter. It is a necessary thing frequently to review our Past conduct as Tradesmen, & to look forward with Fear & caution into the Scenes of Futurity. Things go on very slowly indeed, the late Speculations have done infinite harm to every Article made of Cotton Wool. None of the large Quantity of goods that raised the markets are really sold & we believe will all come back into the market. This will be like the Shock of an Earthquake to many Persons, & will be their ruin—keep snug & quiet—"

S. & W. Salte to S. Oldknow :

" *April* 25*th*, 1787.

" You possibly have done right in purchasing the Estates you mention. We have no doubts arising from it. All we wish is, that nothing may interfere with the Great Pursuit the improvement of the Muslin Manufacture. All the rest should be 'Trifles light as air' in your Consideration. We are poor enough indeed from the long Credit we give & a prompt at the India House last Week for many thousands. What are to become of all the Speculators in Cotton Wool & in Callicos &c—are stopping here dayly—sad times ; goods are coming to town and selling for an Old Song, particularly Callicos. Get out of all this business."

Richard Arkwright, Junior to S. Oldknow :

" LONDON, *May* 19*th*, 1787.

" . . I have heard from several people of the drop my father has made in the stocking yarn and I suppose it is as Mr Burden tells you, but I know nothing for a certainty I can also imagine twist is offered at 15 or even 20 pr ct under price *good enough* for common callicoes, and indeed I do not see any reason why all sorts of twist under 40 or 50 should not be lower but I do think that fine twist above 55 or 60 will not be lower *provided it is good.* My opinion in this is no better nor so good as yours or other manufacturers, for it must depend upon the demand for fine and the incapability of most of the spinners to make the fine twist good—

" It will answer better for most people to take 20 pr ct off under 30 than to spin fine. This is my real opinion and I have often given it you.—I have no reason at present to alter it. Fine cottons have rather advanced since I came here—I deferred purchasing till the last in expectation of its coming down in which I have been deceived. I have givn 2/7½ and 2/8 London tare and 2/7¾ Man'r tare, delivered at Man'r—The advance of fine cottn and the drop in the inferior sorts is a reason (if there is to be any reason) why low numbers should fall in price.

" I suppose there will be bad work amongst some of the speculators, but not so bad as is imagined by some people—I think Mr Salte shows the worst side of matters. They have been very busy. Prints are quite a drug. I do not hear that goods are going to France as was expected. You are certainly a great way from Anderton and whether it would be right to give it up or not I can not pretend to advise."

S. & W. Salte to S. Oldknow :

" LONDON, *May 25th*, 1787.

" We hope you have misrepresented in part the state of your Manufacture. We always cautioned you against doing to much . . . We applaud your resolution of leaving Anderton & contracting your designs . . . the times and the Trade are both very precarious & no man has much encouragement to venture far. We think they will be worse. Everything is over done & in consequence every article sold for loss or no Profit. The commercial Treaty which was to have done such wonders has yet done Nothing. No stir, no bustle, no shipping on that acct. We have sold a small parcel of our Muslins to two young men that are going over to Rouen in Normandy. We flatter ourselves we shall do something. We have also sent a small assortment of them to St Petersburgh . . . Are you about anything new ? Let us know. Whatever hints we can collect, shall do it agst you come to London which will be next month & the India Goods are upon shew."

S. Oldknow to his brother Thomas :

"LONDON, *Saturday, June* 9*th*, 1787.

"I cannot leave this place this evening—on acct of every place being taken in both Mails . . . if I say any thing abt business it will be wor[se] than ever—therefore I must conclude. Tell Wm Bradshaw to give out no more warps—only upon a new regulation of prices—if the weaver does not chuse to take it out let him fetch in his Reed & Healds—let no more warps be sent to Anderton."

Richard Arkwright, Junior to S. Oldknow :

"LEICESTER, *June* 11*th*, 1787.

"I am so far on my way to town, and is it no small disappointment to me, that I did not meet with you either at Derby or Loughboro. I accidentally saw your servant cross the bridge at Mr Strutt's, and desired he would inform you I was obliged to go away from thence early this morning.

"We were flattering ourselves at Mr Strutt's last night you would come there to take a bed, but I suppose you returned by Nottingham. I should have liked very much to have seen you, and to have heard what is going on.—I do not mean to make a long stay in London. I wished to have known how you were in Money matters and I will be obliged to you for a line saying if you can with convenience make me some remittance on my return home, and how much you think will suit you in this month and the next. This information I want for my government in London, and I only wish to have [what] [perfectly] suits you. The coach is waiting. I shall be at Mr Simpson's Distiller, Holborn."

S. Oldknow to Richard Arkwright, Junior (in reply to preceding letter) :

"I think myself unfortunate I did not get a place in

the Mail Coach last Saturday night to have been at Derby on Sunday agreeable to my first plan where I should have had the pleasure of seeing you, and much so we did not meet each other at Loughboro' as our time of passing through that place must be very near.—

" I was in the Leeds Mail and as my horse was at Coates one Mile on the Nottingham Road, I did not get out at Loughborough.—

" [undecipherable] . . . I did not get home till last night and there being no post I could not make a more early reply to your favour from Leicester than the present. With regard to what is going on in London—indeed my good Sir there is nothing going on to any good purpose.—and in regard to money matters, I am poor enough—but as I owe no sums of any consequence to any body but you and your father—I can very conveniently pay you 1000£ on your return and unless your father wants his money I can pay you another 1000£ next month. If he should I will [try] to do it for I have the goods ready for market and I am determined [to make less than I have at present]. There seems to be no want and to force a trade will be to a great loss. I intend to take a few thousand pounds worth to [ ] & try my luck there, and at Manchester I shall open a sale room next week. No doubt but some of these markets enrich one.—besides my outgoings at home shall not be above half what they were a month ago—so that more will be applyed to pay where due— What signifies making goods when nothing is to be got and labour lost into the bargain."

Richard Arkwright, Junior to S. Oldknow :

" Bakewell, *June 21st*, 1787.

" On my return home on Monday by way of Chesterfield I found your favour of the 9th and yesterdays post brought me back from London yours of the 16th. The reason of my not receiving it sooner whilst I was in Town is fully handsomely accounted for.

"The dullness of Trade is the general cry and I have no doubt but it is peculiarly bad. From the large stock of goods on hand I do not see how it can be expected to be better for some time to come. What are the Scotch doing? I have an order for some twist for which a bill was sent £300 (by mistake it went to Cromford) this shews they want twist there, and all we send weekly of the finer numbers is always almost to a bundle disposed of every week's end. How to account for this I do not know, are they making some particular kind of goods? They certainly would not continue a losing trade.

"You may either remit me this month or not, just as it suits you but I only wished to know that I might provide accordingly.—

"Mr Simpson informs me he took off the discount of a parcel too much, when [he] settled the Manchester Mill twist account, but it agrees with the account which T.S. himself gave me, and I think it agreed with your books, but should there be any mistake on our settling it, you will please to adjust it in any succeeding account with them because I have balanced my books agreeable to what I have already received.

"I do think twist will soon be lower, and therefore I would wish you to let your weekly order be as small as is convenient to yourself in just carrying on, because it will be a loss to you to have any stock of twist on hand, without being any advantage to myself, for all our finer numbers will be sold to Glasgow, where the manufacturers take them up as fast as they are sent.

"I perceive by your last letter you have been twice at Derby on your return from town—There is certainly something going on betwixt Miss S[trutt] and you and of a very serious nature too. . . . So I think.

Perhaps you will inform me more when you call here *on your way to Derby* where I presume you are going the next week. As you must on such an occasion be in haste and possibly not able and willing to stay here long, I wish to be informed the day you think you shall come, unless you come in the evening and take a bed with us (which will be at your service) if so, I am certain to be at home in the evenings.

" If you mention the exact time at Derby you will I am sure be then agreeable to it. You are yourself punctual and there is nothing like it.

" We expect Miss S[trutt] here on a visit shortly, the exact time I do not know."

Richard Arkwright, Junior to S. Oldknow :

" BAKEWELL, *September* 6, 1787.

" I hardly suppose this will meet you returned out of Wales. . . . I hope you have seen the Mills at Holywell and will give me an acct. . . . I hope trade is growing better, we want more demand for our middling numbers of twist—Fine and coarse would sell, if we were making those sorts only at present would be the best—I have not yet opened my new warehouse in Manchester, nor have I got one to open, but I will do it when I think it the best opportunity. I am not tired of sending twist to Glasgow, but I have sent too much. I mean to continue sending some. I am not in the way of giving you any news. They say Mr Macmurdo and Miss Strutt are on good terms. I know nothing at all about it. You are very welcome to the use of my horse. I shall be glad to hear you are got well home and that you are likely to use twice the quantity of twist very soon."

Richard Arkwright to S. Oldknow :

" BAKEWELL, *September* 1787.

" Your favour dated 30 August I have received covering 8 bills together One thousand six hundred and seventy four pounds 7/– which balances for the twist in Aug/86 with interest for which I am obliged to you.

"We got safe home on tuesday evening all well—I did not observe any twist was going for you on tuesday last—We must take a discount off it and also what is to be sent you weekly.—I believe 7½ pr ct at present, but that I can not certainly tell, but it will be shewn in the next invoice.—

"I am sorry the weavers have complained of the twist No 60. It should be good for the number, but these lumps ought not to be in it. I hope it will be better. You have not been to Derby at the races.—Nor have I—We have some expectations of seeing Miss Strutt here ere long—then perhaps we may have the pleasure of seeing you also."

The reference to Holywell has an important significance. John Smalley, of Preston, the first partner of the elder Arkwright, had quarrelled with him, and after a period of rivalry at Cromford had removed to Holywell in Flintshire and set up a spinning mill there in 1777. Soon after his death in 1782 a much larger mill had been erected by his sons which, though completed in six weeks, was 40 yards long, 10 yards wide, and six storeys high and was lighted by 198 sash windows "which nightly exhibit a most glorious illumination." It was worked by a water wheel 20 feet high and 7 feet wide with a fall of water of 20 feet. The labour was largely furnished by apprentices, who, about the time of Oldknow's visit had been increased to three hundred. The boys and girls were lodged in separate houses built for the purpose which were whitewashed twice a year and fumigated three times a week with tobacco-smoke, the bedstocks being frequently sprinkled with rectified oil of tar. "A surgeon is appointed to superintend their health, and a Sunday school is regularly attended by a Master at each house." "Our little children [says the employer who supplied these details to the historian in 1795] sleep three in a bed, the larger sizes only two ; and those who work in the night are so far from succeeding each other in the same beds that they do not even sleep in the same rooms."[1] It may easily be imagined with interest these arrangements of a model factory were noted by the new proprietor of the Mellor estate.

In October 1787 Oldknow was up in London at the time of the Autumn Sales of the East India Company, which threw upon an already stagnant market an amount of muslins and calico 60 per cent. in excess of the average importations of the seven previous years.

S. Oldknow to Thos. Oldknow :

"LONDON, *October* 18th, 1787.

"DEAR BROTHER,

"I got here this morning and have been with Mr Salte at the India house—the private Trade Sales comes on to morrow and the Company has declared for a very large Sale the 25th of next Month—no less than Six Ships Cargo—and another private Trade Sale will be in Feby next so that there are more India Goods coming into the Market than has been known of these many Years in so short a time—we must drop making Some Articles—which I will write particularly tomorrow,—it is a pitty we can not get faster forward than we do in the Cuting branch—we shall suffer by old patterns—the waterside house must be got ready and we must push

[1] *History of Whiteford and Holywell*, by T. Pennant, 1796, pp. 214–5. I am indebted for this reference to Mr. A. Seymour Jones, of Wrexham, who kindly placed his collection of materials concerning Arkwright and the spinning inventions at my disposal.

forward the Cuting work there and at Stockport both. . . . A war is expected here."

S. Oldknow to Thos. Oldknow:

" LONDON, *Friday*, *October* 19*th*, 1787.

"I wrote you last night and said that I would write you what Sorts to stop to night—but things are in such a state here that it is very difficult to say what to stop & what to go on with—there is no Trade—Messrs Saltes have not opened the three last packs sent them—so that do not send them any Goods till I get home— Nobody will buy till the India Sale is over—I was in hopes to have discovered an opening for the increase of our weavers but it will be wise to reverse it— in every sort of goods that we have 60 pieces we will not make any more of till that number is reduced to about 10—the Jamdannies we must very much improve in fineness & in taste in patterns —I am told how the Scotch out do us in all respects in these Articles—Jno Livesey & Co has opened their Muslin Warehouse—and they Bost how much they shall out do me—that they have got my man who had the principal Management of the Business—and want to make the world believe that it was owing to him my Goods have been in the repute they have—Many other Makers are sending finer things to market than we make which is rather against us—such as the Goods we have had sent by both houses as a sample—We must make finer goods & we must be doubly carefull in all respects—and not make more but rather fewer than we now do—for it is almost incredible the quantity that will come into the Market of India Muslins the next 4 Months—

" The private Trade sale began to day and very fine thin goods sold high—so that very fine thin goods we must make our aim. Mr A—— sent us some 85 this week and we shall have some still finer the next—do tell

Wm Bradshaw all these things—do push forward the finishing of the Looms—tell Wm he need not do anything to wards fixing another yarn beam till I come home for I shall shew him how—I am desirous of Making such things as no body else is Making and to have them done in our own Shop—so that do push forward the Looms and the finishing waterside house.

"They are all well in Lad lane."

The effect of the competition of India combined with that of Lancashire and Scotland was to give Oldknow a stronger impetus towards the adoption of the factory system, not only in spinning and the finishing processes but also in weaving. On his way North he visited a spinning mill recently started by a former fellow apprentice, John Cowpe, in partnership apparently with one of Oldknow's uncles, at Pleasley in Derbyshire, and immediately after his return he must have set out on a visit of inspection to Dr. Cartwright's new power-loom factory at Doncaster. It was scarcely three years before this that Cartwright, a clergyman and a poet, stirred up by the talk of Manchester men at Matlock about the over-supply of yarn that would follow the lapse of Arkwright's patent, had set himself to invent a "weaving-mill." In his loom as first patented in April 1785 "the reed fell with the weight of at least half a hundredweight, and the springs which threw the shuttle were strong enough to have thrown a Congreve rocket." In short, it required the strength of two powerful men to work the machine at a slow rate and only for a short time. In spite, therefore, of the fact that the machine was so simple and cheap as not to cost more than £5 or £6, it is not surprising that the manufacturers at Manchester where it was made and exhibited did not think it practicable. After two years of improvement covered by two further patents Cartwright felt encouraged to set up at Doncaster a factory of twenty power-looms, ten of which were for muslin, and at the very moment of Oldknow's visit he had just presented a piece of muslin woven by water power to the wife[1] of his friend the Bishop of Durham. The following letter from Salte is unfortunately our sole source of knowledge about one of the most interesting contacts of Oldknow's varied career.

S. Salte to S. Oldknow :

"LONDON, *November* 5, 1787.

"We are pleased your Excursion to Doncaster answered your Expectation. It often happens both Hope and Expectation meet with cheats like common Mortals in their passage through Life. Mr Cartwright was once Professor of Poetry at Oxford & really was a good Poet himself. But it seems he has left the Barren Mountain of Parnassus & the fountain of Helicon, for other mountains & other Vales & Streams in Yorkshire & he has left them to work in the Wild large & open Field of Mechanics—be it so, and may his schemes prosper & fill his purse with Gold. You say not a word about the probability of success, likely to attend his weaving Invention. Can this new Automaton perform the wonders in weaving so Confidently & so flatteringly held out to the World . . . ? Mr. Arkwright was a happy Mechanic. In his Life time he has received the reward of his Ingenuity—It does not happen so in general. We think Mr Cartwright will not be equally fortunate & we cannot give you a reason (from our Ignorance) for this Opinion. You will satisfy us a little more. I wish I could tell you the demand for your Goods was tolerable ; it is otherwise

[1] Baines: *Cotton Manufacture*, p. 230; "Life of Edmund Cartwright," in *D.N.B.*; and *Memoir of Edmund Cartwright, D.D.* (1825).

. . . We are sorry the prospect looks so dark upon British Manufactures—the Private Trade Sale is Over & Goods in general cheap enough but we dread the large Company Sale."

The crisis of 1788 fulfilled Salte's gloomiest forebodings and was an event of the first importance in the history of the Lancashire cotton industry and of Manchester commerce. The failure in April and May of some of the largest London houses connected with the Manchester trade, followed by those of Livesey, Hargreaves & Co., of Blackburn, one of the leading firms of manufacturers, of the oldest bank in Manchester, and of a great multitude of smaller concerns produced strong reactions in public opinion and in the organizations of trade interests. Newspaper paragraphs discussed the general causes of the crisis and traced them back to the peace boom following the American War. Not a few, like Salte, expressed relief that the air was cleared of speculation and monopoly. Statistics were collected, pamphlets written and meetings held to advocate protection by fiscal measures and by corporate organization of Lancashire industrial interests, but against this movement the forces of free enterprize arrayed themselves under the newly unfurled banners of *laissez-faire* and free trade and claimed a decisive victory. But these larger aspects of economic history must be reserved for fuller treatment elsewhere.[1]

Oldknow was naturally a protectionist and took an active part in the Manchester meetings got up to demand the exclusion of Indian calicoes and muslins. But as far as his private affairs were concerned the crisis had been reached in 1787, and the advice offered by Salte in the following letters was sound enough. With the larger

[1] In a volume of studies by Professor Daniels, myself, and others which it is hoped will shortly be published.

effects of the temporary restriction of trade on the organization of his manufacture we shall be concerned in the next chapter.

S. Salte to S. Oldknow :

" LONDON, *February* 27, 1788.

" We have heard a strange Acct of your second meeting at Manchester. The unbelieving Jews as you called them, were not at last to be converted by your Arguments, and the Arm of Flesh was lifted up against them. We again repeat our Sentiments that we cannot see what mode of redress you can apply for to Govermt. Sir Richard Arkwright joins us in Opinion that you must make fewer Goods and make them better & this doctrine we preach to you in particular. Mind your own Manufacture and leave the rest to pursue whatever means they may in their wisdom think best . . . Good muslins & British are selling so very cheap that we are amazed from whence they come, but it is from Lancashire or Cheshire."

S. Salte to S. Oldknow :

" LONDON, *May* 30, 1788.

" Things remain here nearly the same. . . . After all the silly amusement of the Large House in Cheapside and the Banking house in Lawrance Lane, we think they must be Bankrupts. There does not appear a possibility to amuse Creditors much longer—Such multifarious and intricate accts, such drawing and indorsing of bills without End or Reason cannot be easily settled. . . . When will the consequences of this Hurricane be known . . . Curse Ambition and Avarice two Principles that spread ruin and misery . . . We speak not from feeling much loss ourselves, but we really pity many . . . who are reduced from comfortable independence to real want and Poverty. My good friend this should be a warning to every man not to make too great haste to be rich. No more reflections, send us a few . . . [order follows]."

S. Salte to S. Oldknow :

"London, *June*.25, 1788.

" . . . Pray do reduce the number of your weavers. Talk not of Ambition to make a figure as a manufacturer in quantity. We do assure you it is necessary for every man to curtail his Trade, almost to stand still, look about him and collect in his debts. We observe Thwaite's letter. He owes too much money these times to us we want the debt reduced and must have it so We say what we think right when we tell him you are doing but little, and intend making less till times mend, and you must if you write to him again, tell him the same . . [small order follows]. This leisure time you should exert your Genius in making some new patns for a future time. . . . In the sincerity of our hearts we have writ you the above strictures to govern (if possible) your future conduct."

S. Salte to S. Oldknow :

"London, *July* 1788 [?]

" . . . The late hurricanes have overwhelmed all Credit and Trade too. We hear your great Lancashire house is to be in the *Gazette* tomorrow night. . . . You must tell Sir Richard we are much poorer then he is, and you must not draw yet. Very few persons that owe us money pay it and many want help and indulgence as to time, so we must borrow time ourselves. We really hardly know who to trust. Houses are stopping here that were never suspected & we hear the same complaints from different parts of the country. . What a re[volution] . . . three short months have produced in the commercial and manufacturing part of the World. Yet Good may arise out of Evil. But how many thousands suffer by the experiment . . ."

Chapter VII.

EXPANSION AND DECLINE OF THE MUSLIN MANUFACTURE

(*By* G. Taylor *and* G. Unwin)

I.

THE trade depression of 1787–8 influenced the whole direction of Oldknow's manufacture. Hitherto his London agents had been eager to take all he could produce of certain kinds of goods and willing to accept other kinds "on approval." Now that they were unable to give more than very limited orders he was forced to manufacture for stock and then dispose of it on the general market.

In July 1786 S. & W. Salte had taken £3,186 worth of goods out of a total output of £5,397—a proportion of roughly two-thirds. In July 1788 they received £550 out of £3,132—roughly one-sixth. A curious stock or memorandum book for this year contains a list of goods "*wanting places in sales.*" This book shows that Oldknow had, for the first time, a considerable stock in hand of manufactured goods. In a letter of June 1787 he had told Arkwright that whilst he did not "desire to force a sale" he was about to try fresh markets. Before 1788 he had opened a saleroom at Manchester. Shortly afterwards the Day Books show that he was supplying a number of small traders and private customers in Manchester and Stockport from goods in his salerooms.

Trade experienced a complete recovery in 1789. The Day Books show that the sales for April and May 1789 amounted together to nearly £21,000, and the year's sales may be estimated at between £80,000 and £90,000. The extensive correspondence which has been preserved, and the accounts in the Day Books, make clear the scope and dimensions of the business at this period.

Oldknow had acquired a wide reputation for his goods and was now receiving orders from all parts of Great Britain, from France, Holland, and Germany. For some time he had been dealing directly with the more important of the customers of the Saltes, and Parker, Topham & Sowden, although the invoice had passed through the hands of the London merchants. These two firms no longer divided Oldknow's output between them.

In July 1789 Parkers' disposed of two-fifths of the whole production and the Saltes of one-fifth. Oldknow had formed important connections with John Phillips and Haworth & Sons, of Manchester, and Greaves & Hodgson, of London, during the years of depression when the London market had been stagnant.

The following summary of the sales (compiled from the Day Books) shows this development :

	July 1786	July 1788	July 1789
Total Sales	£5397	£3132	£9290
Parker, Topham, London .	1866	1588	4067
S. & W. Salte, „ .	3186	550	2112
Haworth & Sons . . .	—	305	10
John Phillips	—	100	1310
Greaves & Hodgson, London	—	351	292
Arkwright, Cromford . .	20	20	216
Miscellaneous . . .	£325	£218	£1283

An account of the output for this period may be derived from a comparison of the miscellaneous pay-tickets, the fragment of the ledger, and the warping records. The first point is that yarn of far higher counts was being used than in early years. Twist of No. 80 was common now that the mule was producing warp as well as weft, while the ledger shows that weft of over 100 counts was being used. Incidentally the ledger also shows that the weavers had become more specialized, working on the same class of goods and with the same reed throughout long periods. A comparison of these records, arranged in tabular form in the Appendix, shows a significant development in the classes of goods manufactured.

The proportion of calicoes and muslins had changed considerably since the earlier period of Oldknow's business. Whereas muslins had once formed only half of his output, they now represented nine-tenths of his production and even more of his sales. Accompanying this development, the weaving records show that there was a great differentiation of muslin goods and a great output of figured goods.

A clear light is thrown on the state of the manufacture during the final period of decadence, which began in June 1792, by the bundles of pay tickets for the two complete months December 1793 and June 1794. The most striking fact that appears is the shrinkage of the business. Only 180 weavers were working in the service of Oldknow, and it is clear that many of these were only on part time. The production was barely a fifth of what it had been four years earlier. In June 1794 the monthly wage bill was £292 for 500 pieces of cloth. Equally significant is the fact that Oldknow had reverted mainly to calico. Its most popular form, Blue Checks (patterns of which have been preserved), alone represented merely half the total output. Calico now occupied the same position in his business as it had done at the foundation. It is noticeable that the muslins which remained were mainly plain goods. Oldknow no longer devoted his attention to design. The following list of weavers shows their classification at a point near the end of the decline, February 1793.

Com. Call.	60	4/4 & 6/4 Muslins	142
9/8 „	20	Jamdarma	17
Towels & Night Caps	50	Shirting	14
Tablecloths & Doyleys	15	Romals	7
Ticks	7	Bengall Strips	4
		Satten „	4
	152		188

II.

The most important feature of these years is the growth of the factory system in Oldknow's business. The nucleus of a factory existed already in the small staff employed at Anderton and Stockport. Particulars of these workers are given in the early ledger. About twenty in number, they included, beside the manager, the staff employed in attending to the spinners and weavers, with several warpers and finishers. But they differed in no essential way from the assistants employed by clothiers from the 16th century onwards. It seems to have been common to engage several members of each family, the men preparing the yarn for the looms and the women finishing the cloth. Payments in the ledger show that the manufacturer had been dependent on other firms for the bleaching, printing, and dyeing processes down to 1786. For some time Oldknow meditated plans for developing these branches, and thus securing a greater control of the manufacture. The prosperity of 1785–6 enabled him to achieve some of his ambitions by the establishment of the works at Heaton Mersey with his brother Thomas as managing partner.

It is clear from the records preserved that the department was controlled in close co-operation with the Stockport manufacture. Eleven lists of goods remain from hundreds of such daily lists and show how closely the wishes of the original designer could be followed in these important stages. Bleaching and printing were the chief operations pursued at Heaton. At a later date (1791) 47 bleachers (including 15 women) and 37 printers (including 2 women) were employed, with two watchmen to guard the croft. These were skilled arts and the workers commanded comparatively high wages. Several fortnightly wage lists show that the bleachers received from 12/– to 17/– weekly, the printers 12/– to 17/–, while the usual wage for a woman was 5/– weekly.

The works included two plants for dyeing, fully stocked with colouring matters and coppers, as well as a finishing or calendering room.

Sometime in May 1787 Oldknow seems to have organized the processes of warping, sizing, and muslin trimming on a factory basis. It will be noted that these operations, which required but a small outlay in the instruments of production, were the most clearly allied to the central work of weaving. The warping mill was a simple but effective device for labour saving, and had been in common use from the 17th century.

Ten unbound books remain in various states of decay to present incomplete records of 13 warping mills in the period 1787–94. It is clear that Oldknow had, at one time, at least 29 mills, but the records suggest that they were not in continuous use and probably never more than 20 were working together. They display a curious lack of system, suggesting that their management was equally irregular. Many volumes relate only to one mill, but there are considerable gaps in the working. Others have a miscellaneous collection, *e.g.* one volume records the output of Nos. 19, 17, 26, and 29. The earliest book contains an account of mill 19, from March 1787, and the last of No. 7 in November 1794. One man was in charge of each mill, his name being entered daily at the head of each page. The warper rarely worked at one mill for above 12 months. The working week contained only four or five days. A warper received 2/– per day. An interesting series of orders or instructions to workers has been preserved, an example of which is appended below.[1] It will be realized how important it was that the entrepreneur, who alone was in touch with market conditions, should direct this process, for the nature of the warp determined the pattern and quality of the manufactured goods.

It is not surprising to find that the developments in warping were followed by a similar step in the subsequent process of sizing. One of the manuscripts proves that by July 1788 Oldknow had undertaken the sizing of all warps. By relieving the weaver of this task the entrepreneur prevented the delay which sometimes attended the drying of the yarn and secured a greater uniformity in the final appearance of the cloth. The procedure of work is clear. The warp was taken from the " mill " in " ball " or rope form and sent immediately to the sizing house. Here it was passed through tubs of sizing mixture. After impregnation, the yarn was partly dried by a wringer and then placed on a clothes-horse near the stove. From the size house the warp was despatched to the weavers' giving-out room.

In 1788 about 150lbs. of twist were sized daily, amounting to 4,430lbs. in June. As the average weight of a piece length of warp was 3lbs., the output for weaving corresponding to this would be about 1,400 cuts.

[1] *Joshua Battye's Pattern.* (*Mill*) *No.* 10.
68 6/4 Little Cross. Plain Ground.
Pole 204 dents, 12 dents in a half Beer,
24 cord, 2400 double.
To be a cord stripe pattern, 19 – 1 ends.

The finishing processes were of great importance, for the value of the weavers' work depended largely on the final appearance given to the cloth. From the first they had engaged a great deal of Oldknow's attention. His letter in the autumn of 1787 indicates that further developments were proceeding. "It is a Pitty we cannot get further than we do in Cuting branch, we shall suffer by old patterns. The Water-side House must be got ready, and we must push forward the Cuting work there and at Stockport Bottom."

Some records of one of these finishing factories have been preserved. The most important work carried on comprised the trimming of the "float" or figures of the pattern woven, and the cutting of ballasore and romals into handkerchief size. The apparatus of this department was simple, consisting of 30 finishing frames and several darning frames. The staff comprized 81 girls and a foreman or forewoman. The names of the girls suggest that members of one family often worked together. Of the 58 families represented only two or three have names borne by Oldknow's weavers.

A separate time account was made out for each finishing frame, noting the exact time of commencing and finishing work. There were commonly two experienced girls, three improvers, or four learners working on one frame, and the hours appear to have been from 5 a.m. to 7 p.m. The forewoman in charge made a weekly report to Oldknow on the girls and their work. This Disgrace Account, as it is called, contains a number of ciphers, which probably signify the number of mistakes made by the workers, and a report on conduct.

The records also afford brief but clear indications of the gradual transition of the winding process from the domestic to the factory system. Towards the end of 1788 there were about 90 outside winders, who wound in their homes about 70,000 hanks a month. Winding had always been done by children or old people, who earned one or two shillings a week by it. The piecework rates were very low and a refusal to accept them, coupled with a need for a quicker and more efficient service, led to the setting up of nine winding machines in the Hillgate premises, of the operation of which there is a fragmentary record for July and August 1788. The first use made of steam power three years later was for the turning of these machines.

Spinning, with its preparatory processes, and weaving are all that remain to be accounted for. A later chapter will deal more fully with the gradual approach of Oldknow to the adoption of the factory system in spinning. His letter to his brother in October 1787, about pushing forward the looms and having things done "in our own shops," show that he had begun, even earlier, to collect a small body of weavers into his factory. An account for repairs to several looms in 1790 and a record of materials given out to 15 weavers at "S.O.'s Stockport" in 1792, and the fact that a loom-house formed part of the premises offered for sale in 1798, prove conclusively that when William Radcliffe took them over to commence his New System of Manufacture in 1801 the first experimental stage in factory weaving had been passed.

III.

A word must be said in conclusion as to the extent to which the full and detailed accounts of weavers' earnings, preserved in these records, support the glowing account given by William Radcliffe of the great prosperity of this transition period as summed up in the oft-quoted phrase "the golden age of the hand-loom weaver." Radcliffe's account, written in 1822, distinguishes between three periods: (1) before 1770; (2) 1770–88; (3) 1788 and after. In the first period—before the spinning inventions or even the fly-shuttle had come into common use—

> The father of a family would earn from eight shillings to half a guinea at his loom, and his sons if he had one, two, or three alongside of him, six or eight shillings each per week; but the great sheet anchor of all cottages and small farms was the labour attached to the hand wheel . . . it required six to eight hands to prepare and spin yarn . . . for one weaver—this clearly shows the inexhaustible source there was for labour for every person from the age of seven to eighty years . . . to earn their bread, say one or two shillings per week, without going to the parish.[1]

The idyllic effect of this account is marred by the reflection that a father and two sons earning 24/– a week would require a family of at least eighteen—wives, children, and aged parents—to card and spin for them, and that the maximum earnings of such a patriarchal group of twenty-one persons would be 60/– per week.

In the second period—from 1770 to 1788—"cotton was become the most universal material," the hand-wheels were displaced by jennies, the carding for all but the finer yarn was done upon engines, and the fly-shuttle was generally adopted on the looms, which would thus have a greatly increased output even though their number (as Radcliffe thinks was the case) did not itself increase during this period. Of the family income at this time Radcliffe tells us nothing, but though the "inexhaustible source" for infant and aged labour was gone, those engaged in carding and spinning would be earning at least twice the wages of their predecessors The opening of the third epoch, in 1788, was due to "the mule-twist coming into vogue for the warp as well as weft . . . which put all hands in request of every age. . . . New weavers'

[1] Radcliffe: *The Origin of Power Loom Weaving*, pp. 59–60.

cottages with loom-shops rose in every direction, all immediately filled, and when in full work the weekly circulation of money as the price of labour only rose to five times the amount ever before experienced in this sub-division, every family bringing home weekly 40, 60, 80, 100, or even 120 shillings per week."[1]

There are strong *a priori* reasons for treating these figures with critical caution. Writing as a disappointed old man in a time of extreme depression for hand-loom weavers, Radcliffe could scarcely fail to exaggerate the prosperity which had undoubtedly existed in his youth. Whilst our records confirm his recollection of a great boom in textiles for two or three years following 1788, they do not tend to substantiate his retrospective estimate of rise in family incomes, and they show that the boom was succeeded by a severe depression. William Radcliffe, it should be added, was born at Mellor and began his career by supplying warps to Oldknow's weavers.

The Stockport ledger account shows 73 weavers in regular work in 1784 and 115 in 1785, and the average earnings in the November of these two years is 14/7 for weavers with one loom, 26/– for those with two looms, and 30/– for those with three. The taking-in book of 1786–7 enables us to identify the number of looms under the management of each small master. There were 300 weavers, with 475 looms, sub-divided as follows :

With 1 Loom	With 2 Looms	With 3 Looms	With 4 Looms	With 5 Looms	With 6 Looms	With 7 Looms
193	68	19	15	2	2	1
Average earnings of each class :						
13/11	28/7	38/4	48/4	65/10	—	—

Unfortunately there is no record of the same completeness for the two years of exceptional prosperity. A

[1] Radcliffe, *op. cit.*, pp. 61–2.

considerable fragment of a weavers' ledger for July–December 1790 enables us only to ascertain the average earnings of a score of individual looms, a few of which were engaged on calicoes and shirtings and the rest on muslins. One weaver of figured muslins earned as much as 33/– a week, and the earnings of three others were 23/9, 23/1, and 20/3, but the average of all the twenty weavers is only 12/7. A complete set of weavers' pay-tickets for June 1794, when Oldknow was just about to abandon the muslin manufacture representing 180 weavers and 223 looms, shows the average earnings to have been:

1 Loom	2 Looms	3 Looms	4 Looms	
145	27	5	3	weavers
7/1	14/8	19/7½	23/6	earnings per week.

But these figures reflect an abnormally depressed condition of the industry. Of the 145 weavers with only one loom, nearly one-third received from Oldknow less than a pound during the month, but as these only brought in one lot of work in the four weeks they were either suffering from under employment or found work elsewhere. The weekly average of the other two-thirds is 9/1.

If we accept the estimate of Radcliffe for 1770 as a base, our records show at the end of the second period, *i.e.* from 1784 to 1787, an increase of about 50 per cent. in the earnings per loom. They do not substantiate any marked increase in earnings per loom after 1788, but are not inconsistent with a considerable increase in family earnings due to extended employment in 1789–90. This latter indeed may be regarded as certain, but it cannot have been on the scale represented by Radcliffe, and it is very improbable that the earnings of even patriarchal households of two or three families amounted to £6 or even £5 a week.

The weekly wages of weavers of different classes have been calculated for a number of single-loom weavers at a central point of this study :

	March 1787 s.	d.	
Weavers of Fancy Muslins	17	7	Weekly wage
„ „ Romals	16	4	„ „
„ „ Plain Muslins	16	0	„ „
„ „ Aprons	15	6	„ „
„ „ Handkerchief	15	0	„ „
„ „ Calico	12	0	„ „
„ „ Coarse Linens, etc.	9	0	„ „

Note.—At this time the weaving records show the following distribution of the looms :

Calico	91	Plain Muslins	107	Figured Muslin	17
Fustians	42	Apron	72	Romals	23
	——	Check	43	Stripe	12
Total	133	Handkerchiefs	68		——
				Total	342

Chapter VIII.

SPINNING MILLS AT STOCKPORT 1788–91

THE earliest statistics relating to the cotton industry, apart from the various statements put forward in connection with the contest over Arkwright's patents, are those contained in the pamphlet entitled *An Important crisis in the Calico and Muslin Manufacture in Great Britain Explained*, published in 1788, and these were accepted without any doubt as to their accuracy by the contemporary annalists of commerce, but have been subjected to justifiable criticism by the historians of the cotton trade. The pamphleteer estimates the number of cotton spinning-mills run by water-power as 145 and the average cost at £5,000, besides 550 mule frames and 2,007 jennies together with carding engines and billies for the supply of materials to the hand machines, the total value of the whole being estimated at £1,000,000 and the number employed at 112,000.

The writer of this pamphlet had strong motives for exaggerating the amount of capital invested in the rising cotton industry. He was endeavouring to make out a case, during a crisis that had interrupted a boom in the trade, for protection against Indian calicoes and muslins. Ure concludes his summary of the pamphleteer's statements with the remark, " If we take one-half of the above numbers we shall probably be near the truth." Such a criticism may not be too drastic, but it does not help us much to realise the conditions prevailing in the cotton industry at the time when Oldknow commenced business as a spinner. If, as is possible, there were as many as 145 mills at work in the cotton industry it is fairly certain that the immense majority of them were worth less than £5,000 apiece.

The central feature indicated in the many advertisements of small mills at this time is the carding and roving machinery, turned by water-wheel or horse-power, a stove for drying the cotton, and pickers. The accommodation for spinning, where it is mentioned, is generally very limited and often provided separately in cottages or cellars. There is little or no sign at this period of the collection of handloom weavers in the factories, unless the offer in one instance at a factory sale of "25 pair of weavers' looms," along with jennies and carding machinery, is to be regarded as an evidence of this. It is certain, however, that many of the owners or occupiers of these little spinning mills gave out work to weavers and dealt in calico as well as yarn. Thomas Brentnall, who supplied Oldknow with about a thousand pounds' worth of yarn in the year 1786–7, appears in Tunnicliffe's *Survey* as a calico manufacturer. Richard Heys, who left Oldknow's service, as most of his managers did, to set up for himself, carried on with his late employer a two-fold business of this kind of which we have a full year's account. It might also seem that this was the usual procedure for an ambitious beginner, as it divided the risks of this enterprise and opened up two avenues of credit. William Radcliffe's often quoted account of his own beginnings is so much to the point here that it must be again cited. As a child he carded, spun, and wound for his father and brothers, and as soon as he was strong enough he was put to the loom. A few years' savings enabled him to set up as a manufacturer :

> Availing myself of the improvements that came out while I was in my teens . . . with my little savings and a practical knowledge of every process from the cotton-bag to the piece of cloth . . . I was ready to commence business for myself and by the year 1789 I was well established and employed many hands both in spinning and weaving as a master manufacturer.

. . . From 1789 to 1794 my chief business was the sale of muslin warps sized and ready for the loom . . . chiefly to Mr Oldknow . . . Some warps I sent to Paisley and Glasgow. I also manufactured a few muslins myself and had a warehouse myself for my general business.

After the lapse of Arkwright's patents and the invention of the mule and the billy, the beginner with limited capital might confine himself at first to the production of rovings or of yarn, and later on when his capital had increased might add the other branch to his business. Of one of these alternative methods Robert Owen has given us a vivid account in his *Autobiography*. Beginning like Oldknow as an apprentice to the drapery business, he embarked the savings made as an assistant, with a dubious partner, in the unfamiliar but lucrative business of mule-making, and was left in 1789–90 by the desertion of his partner with three mules on his hands together with the requisite reeling and making-up machinery. He rented a large newly-erected factory, sublet most of it on profitable terms, and set three mule-spinners at work on bought rovings in a small part of one of the large rooms. Owen himself reeled the yarn into hanks and made it up into bundles and sold them to a Glasgow yarn agent at a profit of about six pounds a week.

Owen's little concern in Ancoats Lane, which was without power and only employed three men, can scarcely be called a factory except by virtue of the expensive machinery and its master's ambitions. His high profits were due to the demand for fine yarn, which had not yet been met by the application of power to the mule. More usually the beginner sought to utilize the advantages of water power in conjunction with the cheaper labour of women and children. An advertisement in the *Manchester Mercury* on February 6th, 1787, might almost have been put together for the information of the historian :

To cotton and other manufacturers

To be let for not more than 21 years, a large and commodious building formerly a water corn mill containing two water wheels with a dwelling house and about four acres of land 200 yards from Ruthin in Denbigh. The millstream is supplied out of the principal river. Situation . . . will admit of erecting works. . . . The manufacturer may be supplied with plenty of hands at low wages as there are a great number of grown women boys and girls in the town of Ruthin that are out of employ, no manufactory whatever being carried on there at present, and the wages paid to women at hay-harvest does not exceed eightpence a day upon their own meat. The proprietor will advance any sum not exceeding £200 towards fitting up the buildings for the purpose required and will give every other encouragement.

The owner of the Ruthin mill had no doubt been fired by the example of Holywell, in Flintshire, where Smalley, the original partner of Arkwright, had erected a mill in 1777 to which in 1783 there had been added another more than twice the size, a six-storeyed building with 198 sash windows, and a water wheel 20 feet high. But the most interesting feature in this advertisement is the encouragement offered to enterprizing young men with no more capital than Owen or Radcliffe at this time possessed to take out commissions as second-lieutenants of industry. In view of the assistance that was to be had from the corn mill-owner on the one hand and from London firms like that of S. & W. Salte on the other, it is easy to appreciate the force of Radcliffe's statement that "after the practical experience of a few years any young man who was industrious and careful might then, from his earnings as a weaver, lay by sufficient to set him up as a manufacturer"—though but few had the courage. Whilst, however, as far as the future of the industry was concerned, the most significant aspect of this period was the increase in the number of small factory masters, some of whom were to be leaders in the cotton world ten years later, the more striking and immediate development was the increase in the number of the larger factories in which power was applied to the newer types of spinning machinery. Steam began to be employed in cotton mills in 1789 but water power remained for another generation the chief agency, and the competition amongst industrial capitalists with considerable resources to secure water power in the most favourable situations and to develop it by ambitious feats of engineering is one of the most remarkable features of the last decade of the 18th century. The title deeds of Marsland's mills furnish data which will serve as a background for the story of Oldknow's experiments at Stockport and of his more permanent enterprize at Mellor.

In 1786 Marsland secured a confirmation from the lord of the manor of the lease of land and water rights he had taken over three years earlier from the owner of Park silk mills, and in 1789 he acquired a plot of adjacent land of about the same dimensions and built a new factory, fifty-four yards long by twelve, upon it.

By this time most of the larger silk mills in Stockport must, like the Park and Logwood mills, have been converted to cotton spinning. This had certainly happened to the two Carrs silk mills before 1785,[1] and a mill in Adlington Square which was probably the old silk mill

[1] In the *Manchester Mercury* for Sep. 20, 1785 there is announced the sale by auction of the "Carrs Silk Mills now used as a cotton manufactory being 6 stories high 28 yards long and 8 yds wide within with two additional buildings of similar size and several smaller ones and two water wheels one of which has turned 14 carding engines and the other 2 and when a fire engine which is already begun is erected it is expected that as many carding engines will be turned by the latter as by the former."

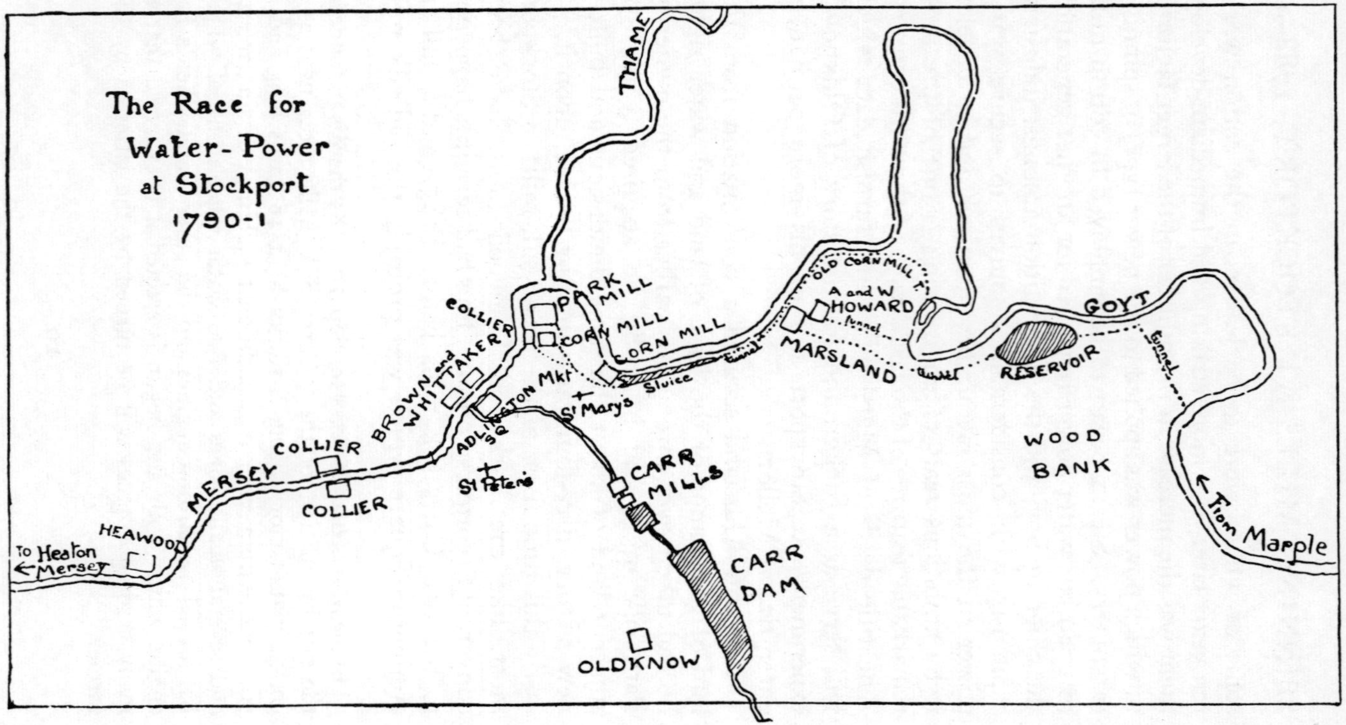
The Race for
Water-Power
at Stockport
1790-1
THAME
PARK MILL
CORN MILL
CORN MILL
COLLIER
BROWN and WHITTAKER
Mkt
St Mary's
Sluice
St Peter's
COLLIER
COLLIER
MERSEY
HEAWOOD
To Heaton Mersey
CARR MILLS
CARR DAM
OLDKNOW
OLD CORN MILL
A and W HOWARD
Tunnel
MARSLAND
Tunnel
RESERVOIR
GOYT
Tunnel
WOOD BANK
From Marple

was being operated by John Collier, its owner in 1790, as a cotton mill. As this resource was getting exhausted many ambitious schemes were afoot for building new cotton mills and for appropriating additional water power. Nearly opposite John Collier's mill in Adlington Square, on the Lancashire side of the Mersey, were two new mills, one belonging to James Brown, the other held in partnership by James Brown and John Whittaker. A little further down the river John Collier had acquired land on both sides, built two new mills and erected a weir between them, whilst further down still at Brinksway Banks, beyond the limits of the town, and on the Lancashire side, Joseph Heawood had a weir and mill.

New mills were rapidly arising in the Park. In April 1790 the indefatigable John Collier secured a 999 years lease of land there at a rent of £5 10*s.* 0*d.* and of water rights at a rent of £100, and erected a factory and weir "below the tunnel-mouth and water-gate of the Stockport Corn Mill."

The lord of the manor had not remained unaffected by this boom. On the plot of land hitherto sacred to the corn mill he had himself built a cotton mill which he let to a spinner. In July 1790 Henry Marsland acquired a third plot of land of about the same size as each of his other two and forming with them a compact rectangle of about sixty yards by fifty in the centre of the Park. Larger plans were shaping themselves in his mind, but it would obviously be unwise to buy more land unless his water rights were secure. These were not affected by the mill building further down the river, but were now threatened by similar enterprize upstream. Abraham and William Howard, who had planned to build a mill on the Goyt near the corn mill sluice, obtained power from Sir George Warren on June 9th, 1791, to construct a new tunnel which cut across a big loop of the river behind the ancient corn mill tunnel, and tapped the water supply at a higher point.

The time had clearly come for Henry Marsland to take action of a larger kind. So far in all leases of water rights by the lord of the manor the prior claims of the corn mill had been carefully reserved, as upon them the lord's monopoly of corn-grinding depended. In Stockport the lord claimed the right to grind even the imported corn of his tenants as well as to bake all their bread, pastry and other victuals. Similar feudal rights continued to be enforced by law far into the 19th century. The people of Wakefield finally got rid of their mill soken in 1854 at a cost of £21,500. But in 1791 such survivals of feudal privilege may well have seemed in no small danger of abolition. No event in the French Revolution had awakened more enthusiasm in the breasts of reformers of other lands than the clean sweep made by the Constituent Assembly on the Fourth of August, 1789. "With louder and louder vivats [says Carlyle], for indeed it is after dinner too, they abolish Tithes, Seigneurial dues, Gabelle, excessive Preservation of Game, nay Privilege, Immunity, Feudalism root and branch ; then appoint a Te Deum for it ; and so, finally disperse about three in the morning, striking the stars with their sublime heads." The spirit of reform in England was not yet overwhelmed by reactionary influences and nowhere were the reformers more ardent or uncompromising than in Stockport. One of the first things recorded by the surviving records of the Court Leet which begin in 1791 is of the nature of a general strike against that feudal institution. Fines were inflicted by the jury on 176 burgesses and 462 tenants for making default of suit and service, and upon five others for refusing to be sworn to perform manorial offices.[1]

[1] Heginbotham I., 167.

It may therefore well have seemed wise to the lord of the manor to dispose of the most unpopular of his privileges to anyone who would make a fair offer. And the bargain he struck on June 12th, 1791, was a favourable one. Henry Marsland was to pay him a lump sum of £2,000 and two perpetual yearly rents of £400 and £100. In return he was to acquire the Stockport Corn Mills together with Mill Dam, Reservoir, and Tunnels and the right to construct further tunnels without prejudice to rights previously granted, and "all tolls, suits of mill, token and mulcture, boons, duties, services and customs belonging to the mill; also the Logwood mill, the cotton mill recently erected by Sir George Warren and a considerable plot of land lying between the properties already acquired by Henry Marsland and the river." The land available for the extension of his business in the Park was thus doubled in quantity and his control over the water power completely assured for the future.

II.

We have now arrived at the turning point of Oldknow's life, the year 1790, in which he built a factory for spinning by steam power at Stockport and laid the foundations for the larger mill at Mellor. Of the conditions and motives that led him into this new enterprize we may be enabled from our fuller knowledge of Oldknow's previous career to form a better judgment than was his contemporary Robert Owen. But Owen's account, though written in his old age half a century later, and though not devoid of a certain animus against Oldknow, who had crossed the path of his earliest ambitions, furnishes the best starting-point for this chapter of our history. Its vigorous outline is confirmed in detail by our business records, and it adds a romantic episode to the story which we could not expect to find in them and which we have no reason to call in question. We shall return more than once to Owen's narrative. Here it will suffice to cite his description of Oldknow's plans in 1790–92.

> Samuel Oldknow . . . had not long before made seventeen thousand pounds of profit in each of two successive years, . . . He had made these profits in the manufacture of muslins, while he purchased the yarn from the cotton spinners. He thought the spinners were getting great profits, and he was not like many others, content to do well or very well as he was doing,—but being ambitious, he desired to become a great cotton spinner, as well as the greatest muslin manufacturer. He built a large, handsome and very imposing cotton mill, amidst grounds well laid out, and the mill was beautifully situated, for he possessed general good taste in these matters. In fact he was preparing and had made great advances to become a first-rate and leading cotton lord. He had however expended his capital so freely in building this mill, fitting it with machinery and purchasing land around it, in addition to splendid buildings and arrangements in and near Stockport for carrying on his extensive muslin manufacture and for its sale, that when the trying time of 1792 arrived, he was too wide in his plans to sustain their expenditure without making great sacrifices.[1]

Undue ambition was certainly an infirmity of Oldknow's mind, but it was not mere greed for further profits that led him to set up as a spinner. The chief problem of his business from the first had been that of ensuring adequate and regular supplies of yarn of increasing fineness at prices that would enable him to compete successfully with other muslin makers whether of India, Scotland, or Lancashire. As early as 1784 he had been recommended by Arkwright to start a spinning mill, and the boom of 1786–7 had led him to acquire land and water-power at Mellor for that purpose. The crisis of the autumn of

[1] *Life of Robert Owen*, p. 55.

1787 had put a stop, for the time, to these plans, but it had furnished an additional motive for resuming them when means were available. The exceptional profits of 1789–90 supplied the means and the incentive to industrial expansion. The factory system in lesser or greater forms was becoming universal in spinning. Oldknow's neighbours, the calico manufacturers of Stockport, were solving the problem of the yarn supply by setting up as factory spinners and were running a race with each other to secure the water-power of the Mersey and the Goyt. Most important of all perhaps was the fact that the finest spinning upon mules was now for the first time being attempted in large factories. Here we may return to Owen's narrative :

> About this period [1790] cotton spinning was so profitable that it began to engage the attention of many parties with capitals. Mr Arkwright . . . had had a cotton spinning mill erected in Manchester, under a manager named Simpson ; and a Mr Drinkwater a rich Manchester manufacturer and foreign merchant had built a mill for finer spinning, and was beginning to fill it with machinery under the superintendence of a Mr George Lee, a very superior scientific person in those days. Mr, afterwards Sir George Philips was desirous of building a large mill in Salford and he unknown to Mr Drinkwater, formed a partnership with Mr George Lee.
>
> Mr Drinkwater being totally ignorant of everything connected with cotton-spinning though a good fustian manufacturer, and a first rate foreign merchant and by this time become very wealthy, was greatly nonplussed by Mr Lee thus abandoning the establishment . . .

That manufacturers should seek to invest their surplus capital in spinning mills was inevitable. The prices of the yarn, which constituted half the cost of their fabrics, had been reduced twenty-five per cent. in a few years by the new machinery, and experience was to show that by the application of power and the organization of labour it could soon be further reduced by a like amount. No doubt, as soon as the new improvements were standardized, the benefit of the economies thus effected would be secured to the manufacturer by the competition of the spinners, but in the meantime he might fear that if he did not secure them for himself he would be defeated by rivals who were more alert. These economies, however, were not to be had by the mere investment of surplus capital. They required a gift for management and a power of mastering technique that were at that time extremely rare. To judge by advertisements, the manager of a factory was often an illiterate workman who received little more than the other hands, and who was attracted by the position of authority and the promise of employment for his wife and children. If a more educated and ambitious young man sought such a post at a nominal salary it was with the intention of learning the business and setting up for himself. Most of Oldknow's managers became prosperous masters, and one of them, John Bentley, succeeded to his position as leading muslin maker in Stockport. The recognition of the vital importance of management marks a new epoch in the rise of the factory system. Robert Owen's account of his interview with Mr. Drinkwater when he applied for the post abandoned by Lee deserves to be classical.

How often do you get drunk in the week? . . . I was never, I said, drunk in my life . . . What salary do you ask? Three hundred a year was my reply. What! Mr Drinkwater said with some surprize . . . I have had this morning I know not how many seeking the situation, and I do not think that all their askings together would amount to what you require. I cannot be governed by what others ask, said I, and I cannot take less. I am now making that sum by my own business . . .

[1] *Life of Robert Owen*, p. 36.

SPINNING MILLS AT STOCKPORT, 1788–91

When I arrived at the mill . . . I found myself at once in the midst of five hundred men women and children who were busily occupied with machinery much of which I had scarcely seen . . . I had to purchase the raw material—to make the machines, for the mill was not nearly filled with machinery—to manufacture the cotton into yarn—to sell it—and to keep the accounts—pay the wages—and in fact to take the whole responsibility for the first fine cotton spinning establishment by machinery that had ever been erected . . . I looked grave—inspected everything very minutely—examined the drawings and calculations of the machinery . . . I continued this silent inspection and superintendence day by day for six weeks saying merely yes or no to the questions of what was to be done or otherwise . . . But by the end of that time I felt . . . ready to give directions in every department. Mr Lee had attained what was then considered an extraordinary degree of fineness, having succeeded in producing . . . one hundred and twenty hanks in the pound. I soon improved the quality of our manufacture.[1]

At the very moment when Robert Owen was grappling with the difficulties of his new position Samuel Oldknow was erecting a mill and machinery for fine spinning at Stockport, in which before September 1791 he was producing yarn as fine as 140 and 150 hanks to the pound. This was not the "large, handsome, and very imposing cotton mill" above referred to by Robert Owen, which was situated at Mellor, in Derbyshire, and though commenced in 1790 was not completed till 1793. The Mellor mill, which produced only comparatively low counts of yarn, was the sole survivor after 1796 of all Oldknow's industrial undertakings and was the centre round which he afterwards re-shaped his ambitions. But in 1790 it was only part of a much larger plan. The Stockport works was to become the organizing centre for

[1] *Life of Robert Owen*, pp. 37–8.

a manufacture of muslins and calicoes which would find employment for a thousand weavers living within a radius of five or six miles, and for nearly a thousand other workers mostly concentrated in five or six factories at different points within the same area. The lower counts of yarn were to be spun by water-power at Mellor in a factory employing over four hundred, mainly women and young people and including fifty to a hundred apprentices, whilst the finer counts would be supplied by some fifty mule spinners in the Stockport factory, which was to have a Boulton and Watt engine, and here too most of the winding, warping, sizing, and finishing would be done by another hundred hands—two-thirds of whom, as of the fifty cardroom workers, would be boys and girls. But two other centres for the finishing processes were to be started—at Bullock's Smithy and Waterside, Disley; and the bleaching and printing were to be carried on at Heaton Mersey, where about ninety workers, mainly men, were employed. The picking or cleaning of the cotton, which found work for another hundred-and-fifty women, was still to be done partly in the homes of the workers, but rooms were to be provided for this process in the factories. The majority of the weavers would also continue to work on their own looms at home, but it was hoped that the loom house at Stockport, where some of the more skilled were working under inspection and control, would serve as the nucleus for the establishment of the factory system in weaving. The warehouse in Manchester would maintain connections with the commercial world of dealers in raw materials and of buyers of piece goods.

The recorded operations of Oldknow as a master spinner in Stockport cover about four years from October 1789 to December 1793, but it is only during the latter half of this period, from September 1791, that we find the spinners at work in Oldknow's own premises in the Hillgate, where a new building must have been erected for the purpose. During the first two years of the period the spinning was carried on in the "Carrs"—about five minutes' walk from the Hillgate premises—where the old silk mills with water-power were now being turned into cotton mills. It was in all probability one of these that Oldknow occupied as a makeshift whilst his own two factories were being erected. The obviously experimental and transitional character of the undertaking makes it more representative of the period and lends it a special interest.

At the "Carrs" shop there were spinning and carding rooms,[1] a warehouse, a putting-out room, a warping room, and a cellar in which cotton could be stored. Of the considerable quantity of cotton now bought by Oldknow some was sent regularly to Anderton to be given out to the small spinners there, some was stored in the Hillgate premises, and a great many bales were delivered to various spinners who had little factories in the neighbourhood of Stockport. But from September 1789 to August 1791 by far the largest portion sent to any one destination was delivered at the Carrs shop, amounting in all to about 380 bags of an average weight of 400 pounds, or some 152,000lbs. Of this total there were still 35,000lbs. left in stock at the end of the two years, so that the rate of consumption was about 6,000lbs. a month. A record exists of "billy slubbing" and low counts of yarn sent up day by day from the Carrs shop to the Hillgate warehouse, and the quantity of yarn sent up in January 1791 was 4,464lbs., valued at £388. Most of the yarn was no

[1] Wheeler's *Manchester Chronicle*, September 24th, 1791, contains an advertisement offering the factory for sale which shows that it contained rooms for spinning—a point left uncertain in the records.

doubt given out to Oldknow's own weavers, but some of it was sold. In the first half of 1791 a London firm was taking on the average about 920lbs. of yarn a month which had almost certainly been produced in the Carrs.

Between fifty and sixty workers were employed there. In the first of the five wage lists that have survived there are 33 spinners, 3 slubbers, and 16 winders, and in the last 39 spinners, 3 jenny slubbers and 15 winders and reelers. There are some new recruits each week and some names drop out, as might be expected, but the great majority of workers are the same week by week. They are of various ages and about equally divided between the sexes. In a list of 37 nine were earning upwards of 10/– a week (the highest wage being 15/–), and these were with one exception males, and three girls earning less than 5/– (the lowest being 4/–). The remaining twenty-five earned between 5/– and 10/–, and the average earnings of all was about 9/– a week. This is a higher average than that of factory spinners at Mellor and elsewhere, and it points to the possibility that some of the spinners were working on their own jennies, though collected into the employer's workshop.

Oldknow probably regarded this small venture as an experiment in mill management which was to be relinquished as soon as his own factories were built and at work. Of the building operations at Stockport we hear nothing, but a record of winding by steam, beginning in March 1791, indicates that a substantial part of the new factory must have been completed in 1790 and the steam engine installed by the commencement of 1791. Boulton and Watt's estimate is worth citing as it indicates the terms on which all the first generation of factory masters obtained their steam power :

	4 horses			10 horses		
All the metal materials except the boiler & the iron-work of the framing of the Engine— about	£225	0	0	£320	0	0
Boiler if of copper. about	40	0	0	90	0	0
Wood framing including its iron-work, uncertain	50	0	0	77	0	0
Putting together likewise uncertain, probably	40	0	0	53	0	0
	£355	0	0	£540	0	0

This is exclusive of all brickwork and all timber of house and floors.

Annual premium to B & W till some time in 1800. 10 horses £50—8 horses £40. but for six horses and under, the premium is paid in one sum viz for 6 horses £210 & for 4 £140 in full.

The letter accompanying the estimate (which has unfortunately lost its date) adds that "it is thought unnecessary to send estimates of engines of intermediate powers as in general they will be nearly proportioned between the 2 estimates to their respective powers." As the premium charged to Oldknow was one of £40, we may conclude that the engine was of 8 h.p. According to Dr. Heginbotham, this was the "first steam engine to be erected in Stockport," and the statement that it was first used to turn Oldknow's winding machines seems to be confirmed by the survival of a record of the operation of ten of these machines, each winding 2,000 to 4,000 hanks per week. Some pages of this book have been lost and the record begins at the end of March 1791. It is said that "the drivers of London coaches when passing the mill slackened their speed in order to tell of the miraculous operations performed therein."[1]

[1] Heginbotham: *History of Stockport* II., 323.

A record of Stock taken September 1st, 1791, reveals the full extent of the changes. In addition to the rooms for winding and warping, cutting and sizing which had been carried on previously on the factory system but without the help of steam power, there was now a carding room with 14 engines, a roving room, and a series of spinning rooms containing 56 machines producing twist and weft in counts from 80 to 150. A fragment of a Time Book gives a complete record for 1792 of all classes of workers in the extended factory except the spinners, most of whom were upon piece work. The 51 warpers and winders, 42 cutters, and 4 sizers who were employed in January 1792 represent the older body of workers, and in addition to these there were 12 smiths and 7 joiners engaged in erecting and repairing the new machinery, 54 carders, and, later on, about a dozen "water-twist" spinners paid time wages, whilst the spinners employed on piece-work, to judge by the record which exists for 1793, must have numbered from thirty to forty, making a total of over two hundred. From a couple of torn pages we learn the weekly wages paid to all the time workers in April 1793. The range of these is remarkably wide. The head smith is paid 21/–, the second 15/–, the third 13/6, four others receive 12/–, two 10/6, two 10/–, and one 9/–. Of the joiners four are paid 18/– and one each 16/–, 11/–, and 9/–. Half-a-dozen carders, one cutter, and two warpers and winders receive wages of from 9/– to 14/–, and the earnings of the rest range from 2/9 to 8/– per week. Two thirds of the whole, we may assume, were young people.

Apart from two more serious stoppages in April and October, the earnings of the spinners must have shown very demoralizing fluctuations throughout the year 1793. Of the actual amounts earned the only evidence forthcoming lies in the two torn pages already referred to, but as these are the records of broken weeks at a time of crisis they are not so conclusive in the case of piece workers as in that of time workers. They seem to indicate average earnings of about 16/– a week.

In regard, however, to the important matter of technique the evidence furnished by the records is very much clearer and leaves us in no doubt of the fact that Oldknow soon failed to keep abreast of the progress made in fine spinning as recorded by Robert Owen in his account of his achievements at Drinkwater's factory.

> In about a year after I had commenced the management of this establishment . . . I had gained the means to increase the fineness of the finished thread from 120 to upwards of 300 hanks in the pound and had thus enabled the Scotch manufacturers in Paisley and its neighbourhood to open a new and extensive manufacture in various kinds of fine muslin, . . . I gave five shillings a pound for the cotton which when finished into fine thread for the muslin weaver, extending to near 250 hanks in the pound I sold for £9 18 6 per pound. This was sold at the commencement of 1792 to Alexander Speirs of Kilbarchan, who made it into muslins, the first piece of which he sent as a present as the greatest curiosity of British manufacture to old Queen Charlotte.[1]

In the accounts of Oldknow's Stocktaking for September 1791, of the daily deliveries from his spinning to his manufacturing departments from December 1791 to December 1793, and of the weekly output of each spinner in his Stockport factory for nearly the whole of 1793 we have an almost complete record of the counts of yarn produced by Oldknow during the period covered by Owen's narrative. The highest count of 150 occurs only in the first of these records ; there are only a few later cases of 140, and even cases of 130 are very exceptional.

[1] *Life of Robert Owen*, pp. 47–8.

The general level of the higher counts is between 90 and 120, and it tends to become lower rather than higher as the record proceeds. Even, therefore, if we make considerable allowance for exaggeration in Owen's account of his own achievements in these years there can still be no doubt that Oldknow's fine spinning did not help him to meet the rivalry of the Scottish muslin manufacturers.

Chapter IX.

THE ESTATE AT MELLOR AND MARPLE

(*By* A. Hulme)

I.

THE old-world villages of Marple and Mellor, intersected by the river Goyt running through a beautiful valley, provided the right atmosphere for the indulgence of Samuel Oldknow's ambition to influence, control, and direct the life of a community by the exercise of a benevolent despotism. This he accomplished by gradually acquiring, either by purchase or exchange, the small estates of the old yeoman stock, and, when established, his dictatorship enabled him to introduce and carry out what to the villagers would be new and revolutionary ideas, not only in the introduction of the factory system, but also in agriculture and road and transport improvements.

His first step in the building up of this new community was the acquisition of the Bottoms Hall estate, which he purchased from Thomas Cheetham and Samuel Cooper. A letter from his solicitor, James Worthington, dated April 18th, 1787, tells us of this purchase :

" I have received a letter from Mr Cooper of Stockport that your Deeds are come back from London properly executed by Mr Cheetham, and I have this day wrote to Mr Cooper that we intend being with him on Monday next the 23rd instant at 10 o'clock. You will therefore be so good as to meet me at the White Lion in Stockport a little before that time, and I think it will be proper for you to see Joel Ratcliffe in the meantime to settle any Difference there may be betwixt you respecting the measure etc, I shall bring £2,000, and you will be so good as to provide the remainder of the purchase money."

Whereas

SOME evil diſpoſed Perſon or Perſons did, on Saturday Night the 3d. of May, or early on Sunday Morning the 4th. wantonly and maliciouſly, break and deſtroy, the leading branches of ſeveral valuable and ornamental

LARCHES,

growing in the plantations on the road ſide, leading from New Marple to Longhouſe-lane, in Mellor.

Any Perſon giving Information of the Offender or Offenders, ſo that he, ſhe, or they, may be proſecuted, ſhall on Conviction, receive

FIVE GUINEAS REWARD,

by applying to Mr. OLDKNOW; and, if two or more be concerned in committing the above offence, any one impeaching the other or others, ſhall be entitled to the ſame Reward.

Mellor, May 5, 1800.

Bancks, Printer, Manchester.

THE ESTATE AT MELLOR

This estate, which is first mentioned in the deeds in 1704, took its name from an old farm known as "Bottoms Hall" on the site of which Oldknow afterwards built his principal farm, which he also used as an apprentice house. That Oldknow had entered into possession of this estate a month before the actual date of conveyance is evident from the following curious letter addressed to him at Stockport :

"MELLOR, *March* 16*th*, 1787.

"SIR,

"Hampsons at Bottams as rid or grubed up 30 trees in the Orchard Aple trees and plumtrees and to day the have rid up a Cherry tree which is things as I supose is Verrey Wrong and what is not Custam as Ever Any farmers used to do. I could be glad to see your Brother over if it will be Convenant Either friday or Seterday as I intend not to Com hom till Seterday Night, from Your Servt

"JAMES GOULDEN"

To Oldknow, Bottoms was the most valuable portion of his estate, as it was on this land that he erected his spinning factory, diverted the river Goyt, and constructed dams and reservoirs (with the exception of the Linnen Clough reservoir) for the provision of water-power. During the same year (1787) he had bought some cottages known as the "Chapel Houses" from Robert Collier; and again on May 2nd, 1788, and November 18th, 1789, he purchased from this representative of an ancient Marple family other portions of the Chapel House estate. It was not till May 1791, however, that he obtained the whole of Collier's Marple estate, including the Chapel House farm (also known as Collier Fold), which he effected by means of an exchange by which Collier entered into possession of land at Bosden which had probably been the site of an earlier factory. It was upon Collier's land that Oldknow built his lime-kilns and sank his coal-shafts. As far as can be ascertained its boundary extended on the north side to where Posset Bridge now stands, and on the south side to the small plantation known as Picky Wood, about 200 yards beyond the Brick Bridge in the direction of Strines. The estate included the Chapel Houses on the westerly side, and in the east must have had the river Goyt as the boundary, the land running parallel with the Bottoms Hall estate, the river between. The evidence for this occurs in a letter dated February 14th, 1787, from George Worthington, an Altrincham solicitor acting on Oldknow's behalf:

"If you mean to purchase the Water, Wood and Land of Mr Collier you mention you will do well to be well informed of Mr Collier's Right over the Water & whether supposing he has & you purchase it, you will then have power to Dam up the Water as you wish without making further purchases of those upon whom you may dam it."

As the names of the meadows and fields of this small estate shewn in the deeds are of considerable local interest, we make no apology for recording them:—Calf Croft, Earkensough Meadow, Earkensough Field, the Further Priest Field, the Nearer Priest Field, the Fitch Croft, the Old Meadow, the Rough Croft, the Hollins of the Rill, the Small Calf Croft, the Field at the Door, the Round Meadow, the Bardsley Meadow, the Long Rangland, the Bridge Hey, the Foxlow Meadow, the Foxholes, the Long Intack, the Ryefield, the Higher Dean Bottom, the Haddon Close, and the Wood.

The Ryefield, before the construction of the canal, covered that portion of land now occupied by the four top locks, the three canal ponds, Jink's boat building yard, and the lime kilns. The Foxlow Meadow situated below extended from the Ryefield down to the edge of the valley, the Old Meadow abutting on the south side now being divided into two fields. The Foxholes, the Nicho Meadow, and the Haddon Close we have been unable to identify. The latter name, however, suggests an ancient enclosure of the one-time lords of the Manor, the Vernons of Haddon Hall.

In the same month in which he made the exchange with Collier, Oldknow also purchased the adjoining estate known as "Goit Cliff Torr" from Joseph Lowe. This land, which adjoins the "Chapel House" estate on the south side, had been put up for auction on January 19th, 1785, at the Sun Inn in Stockport and had evidently been withdrawn, is described in the *Manchester Mercury*, January 4th, 1785, as follows:

Now in the Possession of Mr Joseph Lowe All that Ancient Estate called Goit Cliff Torr consisting of a messuage or Farm House, Barn, Stable, Shippen & other necessary and convenient outbuildings and several Closes, Fields or Parcels of Land containing 38 acres 1 Rood 1 Perch of the Statute Measure or thereabouts. The situation is rural and pleasant having several agreeable objects in Prospect as Wood, Water, etc. The land itself is naturally good and fertile and now in rest and condition, is well known to be the first Estate in the Neighbourhood for growing Timber, and there is now standing thereon, 266 heavy full grown Oaks, fitting for ship or any other use.—56 Cyphers, 144 Ashes, 51 Cyphers and 114 Owlers and Birches and 19 Cyphers and an innumerable quantity of young plants. There is also a coal mine therein which now lies dry and might be got at a small expence, it has likewise the sole Right to the powerful river Goit running along the Easterly side thereof which might be easily appropriated for the use of Cotton or other Manufacturies worked by water and for which buildings might be erected at a moderate charge, there being Stone and Clay for making Bricks therein.

It is clear that Oldknow was not satisfied with Collier's

claim to the Water Rights, and therefore took the advice contained in George Worthington's letter of February 14th, 1787, and made this further purchase (Goyt Cliff Torr) higher up the river in order to secure the right of erecting a weir, the settlement of this question now being vital as the building of the factory had already commenced. In the meantime he had been in negotiation with the Trustees of Thomas Bradshawe Isherwood respecting an exchange of a portion of his Stockport estate for the Isherwood lands adjoining Bottoms Hall estate, and on June 17th, 1791, the following were conveyed to him : Linnen Clough, Hill Top, Lower and Higher Windybottom, and Greenclough Farm.

This exchange extended his holdings along the east bank of the Goyt as far as Strines Hall [1] and up the hillside above Bottoms Hall some little distance beyond the site now occupied by the mill pond which he constructed in 1809. Six months after the exchange with the Isherwood Trustees Oldknow mortgaged the whole of his Marple and Mellor estates and part of his Stockport property to Richard Arkwright for £12,000, on which sum he was to pay interest at five per cent. Whilst he was being driven to this method of raising money to carry on the building of the factory he still continued to purchase farms and lands contiguous to his own, and in May 1792 he acquired from John Fearnley several fields and a farm house known as " Drake Carr " situated at the south-east end of his estate close by Greenclough farm. Adjoining Drake Carr is a small farm known as " Broad Carr," which was purchased from the Turner family in October of the same year.

Subsequent additions were Golden Fields, situated close by Marple Church, purchased from Ralph Bridge in 1818 ; on the Mellor side Miss Shaw's estate, and

[1] Site of Strines Print Works.

in 1823 the Tanyard cottage and the surrounding fields were added.

The Marple and Mellor estates included practically all the land lying to the south of what is now Mellor Road and Longhurst Lane[1] as far up as the little Methodist Chapel close by the Devonshire Arms, then across Cobden including a number of fields lying to the east of Gibb Lane and taking in Coldwall, Primrose Lane being the extreme edge of his land in the east, as far as the remains of the old cross. The holding on Black Lane only included the small farm and land known as "Paradise." Passing Tomlinson's land and Bull Hill farm and following the old pack-horse road down the hillside, we again strike Oldknow's land close by the quarry above the Banks farm. The boundary then takes a turn to the south and includes Cross Gates, Broad Carr, Drake Carr, and a distance of roughly three hundred yards beyond Greenclough farm It then followed a course along the east bank of the Goyt in a northerly direction to Windybottom Bridge.[2] Between 200 and 300 yards beyond this bridge on the west side of the river we strike the Marple boundary, which extends up the side of the hill to a point below Hill Top farm, then along the hillside including several fields above the canal as far as Picky Wood. The Marple estate included the wood on the south side of the Church, and, crossing the Marple to Disley main road, took in several fields adjacent to the road on the west side in the direction of the village. Skirting the "Hollies" on the east, the boundary ran parallel with the canal locks as far as Brabyns Bridge, and included nearly all the land on the south east side of the road in the direction of Low Marple to a point

[1] Originally called Longhouse Lane.

[2] The name Roman Bridge is a modern innovation. It was known as Windybottom Bridge in Oldknow's day.

some little distance below where the railway now lies. With the exception of a few small outlying estates—such as Routing Walls in Marple, Waterside, Disley, and Podmore Farm in Mellor—the above is a sufficiently accurate outline of the boundary of Oldknow's land.

II.

A common practice in this neighbourhood at that period was for farmers to set apart large pastures for the taking in of "Summerlings" or "Joist" cattle as they were called, at fixed prices. Gentlemen's parks were also used in many instances for this purpose. At the time Oldknow took over the Bottoms Hall estate he had not yet purchased sufficient cattle to stock his farm, and he adopted this practice during the summer of 1787.

During February, March, and April advertisements appeared in Wheeler's *Manchester Chronicle* calling attention to the fact that Ley or Summer pastures were to be had on the Bottoms Hall estate. In July of the same year ten men with rates of pay varying as follows: 1/4, 1/6, and 1/8 per day, and five women at 6*d.* per day, are recorded as hay-making. Then follows a gap of three months, and in October and November we find a number of local farmers ploughing the land on this estate and charging Oldknow with the cost.

During the month of May 1788 large quantities of potatoes were set in the ground which had been broken up by the plough the previous winter. The wages paid to male workers for potato setting was 1/4 per day, and 6*d.* a horse-load was paid to female workers for cutting the potato sets to the required size. A significant document of this period (1787–8) is a bill for the making and repairing of agricultural machinery, carts, wheelbarrows, and the shoeing of horses, due to a smith named William

Wood for work done at Bottoms Hall. This account contains 116 items from April 1787 (the month the estate was conveyed to Oldknow) to May 1788, and the fact that most of them are for repairing ploughs and harrows suggests that meadows and pastures were being broken up and turned into arable land.

In the spring of 1788 Oldknow commenced improvements on a portion of the Chapel House estate he had purchased from Robert Collier and planted three thousand "Quick Sets," for which he paid Benjamin Moors 10*d.* a hundred.

A series of letters written to Oldknow by his stepfather John Clayton, who had evidently undertaken to stock the newly acquired farm, not only shed light on cattle fairs and cattle-raising of the period but also afford a glimpse of Oldknow's stepbrothers, one of whom—the younger John Clayton—ultimately succeeded to the management of Mellor mill :

John Clayton to S. Oldknow :

"Roscow Low, *November 4th*, 1787.

"I receivd yours and how to Advise you about Either Scotch Bullocks or Heifers I am at a loss. The reason I have they are sold at Unaccountable prices as there will be Little to be had for keep. However as I am going to some fairs I will buy something as will pay for keep Either Bulls Heifers or Young Barron Cows as you may sell again at spring. If you would keep them all Summer and your pasture would make them fatt, I would advise you to buy some Young Cows or else some Good Scots, Either you or your Brother must buy Sam a Rockin Cheer which he is very fond of and Jenny says he must have One Against he comes home.

"P.S.—I am going to Kendall this week with some wintering Cows. If I can buy any Scots will send them."

John Clayton to S. Oldknow:

"[Undated.]

"I heard of a Cow near Preston and went Over Yesterday. She is 7 or 8 years of age, not to calve of a month; the man knows nothing of her milking. However I bought her. He is to keep her 2 weeks in gel . . . and she cost 11 guineas. She is a good brandle cow and promising for a milker. If you could wait time of her calving she might come to suit you. If not, send by John Bentley and I shall endeavour to buy another. They are very bad to light of at this time and sold at very great prices. My son John wrote to me by John Bentley which I was glad to see. He pleased your mother and me very much by saying he was gott into Division in 4 weeks and has a very good heart of Arithmetick. Wishd us to send the Fife which he says in his letter you were so kind to give him. We have sent it in the weft which is in coming. We are of opinion it will not do for him to blow a great deal upon it as he is not of the Stoutest sort."

John Clayton to S. Oldknow:

December 22, 1787.

"Your Brother Mentioned in his Letter you had a Desire of knowing the prices of the Heffers sent to you which I Have given an Accompt as Under:

1 Brandle fincht Heffer from Warrington .	£6	6	0
1 Red Loophorn from do.	£6	6	0
1 Barren from do. a good deal of White .	£5	15	0
1 Little Heffer from Clitherow Bull . . .	£6	0	0
1 Brandled Heffer Down Horns from Do. .	£5	10	0
1 Black fincht Heffer from Do.	£6	3	6
1 Little Heffer from Do.	£4	12	6
1 White Stirk from Do.	£4	12	6

"These are the prices they Cost Exclusive of the Charges. I purpose Coming either first or second week in Christmas if your Mother Comes Back on Horseback it will be in the Second Week So that I can Bring two Horses and it will Save More Trouble."

MELLOR MILL AND RESERVOIR

Chapter X.

THE CRISIS OF 1792–3
OLDKNOW AND DRINKWATER

FOR the sake of clearness the story of Oldknow's great venture in factory building has been so far told without much reference to the financial difficulties that arose out of it and which became acute before it reached completion. Some account of these difficulties must now be given as they supply the essential background to the new industrial régime of Oldknow at Mellor during its first few years. Not only was the mill at Mellor largely built and run on credit, but the credit was of a kind which, if not altogether new, has never been adequately described. The story of the crisis of 1791–2 is the necessary prelude to that of the Shop-Notes of 1793–4, and it likewise provides the main clue to the rest of Oldknow's career as a Captain of Industry. It brings us back once more to the romantic narrative of Robert Owen, whose substantial accuracy there are no sufficient reasons to doubt, but whose meagre account of Oldknow's business difficulties we are able to supplement from the most authentic sources—the letters of his bankers, his other creditors, and of those to whom he applied in vain for assistance. A letter from his old friend Salte falls under the last of these categories. It was written a fortnight before the close of 1790, the second of Oldknow's two years of unexampled prosperity. He had just built his new spinning factory at Stockport equipped with the Boulton and Watt engine and was laying the foundations of his larger mill at Mellor

S. & W. Salte to S. Oldknow :

"LONDON, *December* 17, 1790.

"We recd yours of the 14th relative to your Estates & mortgages &c. You must do as well as you can being got into the Wood. But to make short of this you are the best Judge of what you have been doing at Mellor or elsewhere & I hope you have not made a bad bargain—Respecting the £5000 upon a mortgage of your premises at Stockport, it will not be possible for us to advance you the sum till near Feby next. We have some unexpected engagements for our money happened very lately, besides prior engagements to yours; another thing the present India Sales runs away with all our Spare Money & more than we expected. To this we may add some disappointment in the receipt of near £6000 which we should have recd before Jan'y but that is now delayed for some weeks—these things all put together put it out of our power to do this business at present—but you have resources in plenty at Manchester without foreign aid, at least for a short time. We want nothing but what we write for on this side the holydays as we shall take stock the last day of this month."

S. & W. Salte to S. Oldknow:

"LONDON, *January* 5, 1791.

"We suppose your engagements at this season of the year have monopolised all your time—so that we have hardly been indulged with a single line. You sent us two ps. more Dimithys, but are returned, as was obliged to furnish ourselves before their arrival. The upholsterer waited for them, and we need not tell you we got them 6d. a yard cheaper in London of equal quality . . . Gardiner was here this day If his order for the muslins for printing do not come next week he will not take them. Do leave Mellor awhile and look after the manufacture. It might be advantageous to you. Pray let us know what you are about and that soon."

From further correspondence and other sources we know what Oldknow was about and need not be surprized at his friend Salte's impatience. By exchange and by extensive purchases he was becoming, during the first half of 1791, a large landowner in Mellor and Marple; and to provide means for this object he was straining every resource in Manchester and elsewhere. The first person to whom he turned for assistance was Arkwright, to whom the large debt for yarn which Oldknow had incurred in 1787 had not yet been wholly repaid. Letters from Richard Arkwright dated July 23rd and 30th, 1791, acknowledge the receipt of two remittances of £400 each on two successive weeks which are apparently part of a system of regular instalments towards the arrears of Oldknow's yarn account, and a letter of September 13th, 1791, acknowledging the receipt of a further £700, complains that these remittances have become irregular. "It is true I did not expect a regular sum to be remitted me every week, but I understood the defalcations of one week were to be made up on others, which you have not done." A similar acknowledgment of £480 "on the Twist Acct." on July 7th, 1792, seems to imply a scheme of payment at the rate of some £20,000 a year for yarn. As it is scarcely likely that Oldknow was receiving yarn from Arkwright at this rate in 1791–2, we must conclude that the prosperous years of 1789–90 had not enabled him to pay off his former debt. But on April 13th, 1791, he had undertaken to mortgage to Arkwright the estates he was then engaged in purchasing as security for a loan of £12,000, which seems to be the first of a series of loans of similar amount which supplied the bulk of the capital for the new enterprize.

In the meantime he was looking for additional help in other directions, and the following letters of Henry Norris show us on what terms a retired merchant was prepared to invest his capital in industry. In 1788, when Byrom's Bank suspended payment, Davy Hulme Hall, the residence of William Allen, the surviving partner in that concern, was bought by Henry Norris, who retired shortly afterwards from his business as a cotton merchant in Garret Lane, though he had still thirty years of life before him.[1]

Henry Norris to S. Oldknow :

" BLACKPOOL, NEAR PRESTON, *April* 6, 1791.

" Since I last wrote you on money matters, I have been in expectation of your having finished your purchase with Mr Worsley, as I have kept myself provided as I advised you. I should be glad to know if you have effected your purpose with said Gentleman."

Henry Norris to S. Oldknow :

DAVY HULME HALL, NEAR MANCHESTER,
June 30, 1791.

" Since I last wrote you from Blackpool, I am not favoured with any of your Letters. In that Letter I noted to you that I had not, nor was it my intention, to place any money at Interest on personal security, except on double Bonds & the names living. I well recollect your saying to me in your last Letter that you would not ask any person to be joined security with you, but I was in hopes you wd change that resolution—It is not my wish to urge you to any measure unpleasant to you, but at same time I dare say you will admit that I have a consistent right to lay down such rules for the conduct of my private affairs as I shall judge prudent—. In this instance it is fortunate the circumstance happens to a person whose credit is above conjecture. I hope you will do me the favour to give me your answer to this Letter."

[1] L. Grindon : *Manchester Banks and Bankers*, p. 48.

Henry Norris to S. Oldknow:

"DAVY HULME HALL, *June* 4, 1792.

"I am favord with your Letter of the 1st in answer to which I am ready to receive the security you proposed, but to prevent any mistake from your Letter I take the liberty to repeat that I will not place out to interest one Guinea but at 5 per Ct."

Apart from the probably ironical reference to the state of Oldknow's credit there is nothing in these letters to indicate the serious crisis in his affairs which must have been reached by Midsummer 1792. At that date he had relinquished none of his earlier enterprizes with the exception of the spinning at Carrs Shop. The wages paid at Anderton amounted in the spring of 1792 to about £50 a week. The weavers of the Stockport district can hardly have received less than £200 per week, and the accounts of the small master spinners may well have amounted to another £100. The wages of the 200 workers in the Hillgate factory, together with the many outside pickers, etc., cannot have come to less than £100, and there are records of the wages at Heaton Mersey Print and Bleach works pointing to an average of about £50. To this roughly estimated total of £500 invested in more or less saleable goods, or in the materials for their manufacture, there must be added a wages bill at Mellor of between £150 and £200 a week, which was almost wholly an investment in fixed capital on which the returns must be distant and precarious. Given an efficient management and an expanding market, this extension of Oldknow's business would have presented no just cause for anxiety. But we have had sufficient evidence that the muslin manufacture which was the vital feature of the enterprize was losing its hold on the market, and in 1791 the death of Oldknow's brother and partner, who had managed the bleaching and printing department, must have seriously diminished the efficiency of the manufacture. That under these circumstances Oldknow should seek a partner who would bring an accession of capital [and engineering skill] into the business is not surprising. [During 1792 the brilliant young engineer Peter Ewart became a partner in Oldknow's firm.[1]]

We need not be astonished at the somewhat curt and sardonic terms in which Arkwright received the announcement of this new development :

Richard Arkwright (the younger) to S. Oldknow :

" BAKEWELL, *June* 27, 1792.

" Sir,

I am favoured with your letter of the — inst covering 4 bills together Two hundred and seventy pounds which is in full for half a year's interest on the £12,000 due 30 inst. The Bill for Mr Mossley I have given to him and he will advise you of the rec't. I have informed him what you say and he will answer for himself.

" I hope the person, whoever he is, that you have joined has £20,000 to spare and will turn out in other respects an agreeable partner."

[The Oldknow–Ewart partnership was not[2] the one which had such a disturbing effect on the career of Robert Owen, for it was dissolved in the course of 1793.[3]] Shortly after Owen's engagement as manager in 1792

[[1] For Ewart (1767–1842), see *Dictionary of National Biography* ; H. W. Dickinson and Rhys Jenkins, *James Watt and the Steam Engine* (1927), pp. 288-9]

[[2] As Unwin assumed it was.]

[[3] See W. H. Chaloner, " Robert Owen, Peter Drinkwater and the Early Factory System in Manchester, 1788–1800," *Bulletin of the John Rylands Library*, vol. 37, 1954–5, pp. 92–98.]

his employer, Peter Drinkwater, sent for him and offered to retain him by a written agreement by which Owen was to be admitted a partner in 1795 along with Drinkwater's two sons, with a fourth of the profits. [Owen's account suggests that his memory had ante-dated the Oldknow–Drinkwater affair by at least two or perhaps three years, i.e. from 1792 to about 1794] :

" I continued happily progressing in my situation [says Owen] until the disastrous year of commerce 1792, checked the rapid upward progress of the cotton trade. All engaged in it suffered more or less. Many were ruined in various businesses all over the kingdom. Mr Drinkwater had a large capital, and stood the shock of this revulsion in the commercial world without much loss . . . Miss Drinkwater was attaining womanhood . . . and had been well and carefully educated . . . A gentleman whose name I forget—a merchant in the foreign trade of Manchester . . . paid his addresses to her. He was not disliked by the young lady, although he was perhaps ten or twelve years older than she was ; for his temper was good and his manners attractive to the other sex . . . In this case the course of true love did not run smooth. I must now recur to a name mentioned in an earlier part of this narrative—the name of one who stood high in Cheshire and Lancashire as the first and most extensive manufacturer of British muslins . . . This was the once celebrated and most enterprising Samuel Oldknow . . . who was then supposed to be very wealthy, and was considered a great man in the world of manufactures and commerce . . . In fact he was preparing and had made great advances to become a first-rate and leading " cotton lord." He had however expended his capital so freely in building this mill, fitting it with machinery and purchasing land around it, in addition to splendid buildings and arrangements, in and near to Stockport for carrying on his extensive muslin manufacture and for its sale, that when the trying time of 1792 arrived, he was too wide in his plans to sustain their expenditure without making great sacrifices. To prevent this it was afterwards generally thought that he considered an union with Miss Drinkwater would, by the assistance of her father, enable him to proceed unchecked. He was a hearty, healthy, handsome man, but yet perhaps five years older, than Miss Drinkwater's present suitor. But I suppose he concluded that 'faint heart never won fair lady'—and therefore he at once applied to Mr Drinkwater to be permitted to pay his addresses to his daughter. Mr Drinkwater was flattered by this application, for at this time Mr Oldknow stood prominent in the cotton world, next to the Arkwrights and the Strutts of Derbyshire.

"It was believed, that he was not well received at first by the young lady. But the father, although fond of his sons and daughters, was ambitious and obstinate, and succeeded at length, by the great prospects held out to her of great wealth and station, in overcoming her reluctance. She had not the means to resist his authority, having no property in her own right, and she therefore yielded to her father's wishes. Mr Oldknow was consequently received and accepted by father and daughter as the future husband of the latter.

"For some time all matters seemed to proceed successfully with Mr. Oldknow and for a certain period he had a great influence over Mr Drinkwater. During this period he became acquainted with Mr. Drinkwater's engagement with me, and this stood now in the way of his (Mr. Oldknow's) ambition. He expressed a great desire that the whole business of both houses should be kept entirely to themselves, and my partnership, which, according to my engagement, was to commence the next year, stood in the way of this exclusive dealing with Mr. Drinkwater's property.

"Mr. Oldknow thought that this difficulty should be overcome at any cost, and it was determined to try what could be done with me. Mr. Drinkwater, who for three years had once only been to his factory in Manchester, and then to introduce the celebrated Herschel the astronomer, sent for me to Newal House,[1] his country residence, where I had not been since the day the agreement of his own proposal was made . . . He said—'I have sent for you to explain unexpected changes which have taken place in my family. The celebrated Mr Oldknow is to become my son-in-law . . . He has expressed a strong wish that the entire business of both houses should be retained in the family . . . If you will give up your claim to the partnership you may name your own salary . . .' He appeared very anxious to hear my reply. I said 'I have brought the agreement with me, and here it is, and I now put it into the fire, because I never will connect myself with any parties who are not desirous to be united with me; but under these circumstances I cannot remain your manager with any salary you can give.' And the agreement was consumed before him."[2]

Owen subsequently agreed to stay with Mr. Drinkwater till he could find a worthy successor for him and did not leave him till nearly a year afterwards. In his later account of his settlement at New Lanark in 1800 Owen says:

"Application was now made to me ... by the Mr. Humphreys whom I had recommended to Mr Drinkwater to succeed me. Mr Humphreys could not keep up to the

[[1] This is an error for Irwell House, Agecroft.]
[2] *Life of Robert Owen*, pp. 53–8.

quality of yarns or to the general management of the establishment as I had left it to him. Mr Drinkwater had discovered that Mr Oldknow's pecuniary position was not what he had anticipated and therefore the match between Mr Oldknow and Miss Drinkwater did not take place. Mr Drinkwater became dissatisfied with the business, sold the factory and Mr Humphreys lost his situation."[1]

Oldknow's ill-fated engagement with Miss Drinkwater was [not] broken off during the crisis of 1792–3, and it seems that her father's benevolent interest in Oldknow's affairs [continued until as late as 1794–5]. The crisis which had its beginnings in the autumn of 1792 and reached its culmination after the declaration of war with France in February 1793, cannot rightly bear the blame for Oldknow's subsequent difficulties. With his affairs already in so dubious a state it is remarkable that he should have contrived to weather the storm, and in fact he only succeeded in doing so by jettisoning one after another all the branches of his enterprize but the latest. The first to be abandoned was the Bleach and Print works at Heaton Mersey, which were taken over in November 1793 by Robert Parker. The Stockport manufacturing and spinning with its headquarters in the Hillgate went on for another year but is not further heard of after the end of 1794, and by this time the Anderton branch also had probably ceased to exist. His main anxiety during these two years of trial must have been to secure the survival of the mill and estate at Mellor, and that he achieved this was due to the forbearance of his creditors small and great. The foremost of these was Arkwright, who appears to have lent another £11,000 in September 1792 and who was not unnaturally annoyed by Oldknow giving priority to the claims of more recent and more

[1] *Life of Robert Owen*, pp. 81–2.

importunate creditors amongst whom were the bankers Smith, Payne and Smith. The following letter seems to mark a crisis in Oldknow's affairs and probably induced him, if he had not already so decided, to abandon all his other schemes and to retire in good order upon Mellor.

Richard Arkwright to S. Oldknow :

" CROMFORD, *May* 13, 1794.

" DEAR SIR,

" I am greatly surprised indeed by your letter and I know not how to answer it. I have several times intended writing to you both with regard to the half years interest due and with regard to a part of the money being paid in. If Messrs Smith are solicitous to have the money for which they have a security, what ought I to be who have *no security*—and for more than double the sum—You have repeatedly told me that you would lessen it and I thought in a considerable degree. The first time I expected a considerable sum was when you borrowed of Smiths, but I soon found you had it not for that purpose. Lately I expected I was to have the money to come in for the sale of your printing works, but I hear nothing of that.—You now propose to pay neither principal nor interest but that I should sit still and let Smiths receive 250£ or 500£ monthly, whilst I who ought to receive three times that sum am to have nothing, but we are to consolidate . . . to eternity . . .

" I have given you every reasonable time and notice repeatedly that I wished to have the money. You know what was my opinion from the first of your launching out into these extremes and you have pretended to consult Mr Strutt & myself what was best to be done, but instead of following our advice you have continued to increase your machinery. I should be happy to render you every service in my power *consistant with prudence and that duty which I owe to myself*. . . . To Mr D—— say . . . Will Mr D—— join you in security to me for the debt and engage . . . the interest shall be regularly paid ? One may *talk* of letting matters rest till 6 or 12 months after the war, but surely such a thing does not deserve a serious thought and is quite unmercantile Probably you may not have asked Mr D—— if he will give security with you, if the matter . . . rest To be conclusive I neither can nor will suffer Messrs S to receive this money whilst my debt is to remain and even the interest unpaid. State one account as it now stands and you will astonish the man you shew it to ; to think that I could rest so long.

" I hope you will devise some means, I can think of none, to get within a reasonable . . . Had you paid the two half years interest on the Mortgage and on the Bond half a year each I should not have been so much disappointed as I now am by the receipt of your letter.—I have no intention at present of going into Lancashire."

Chapter XI.

COMMUNITY BUILDING AT MARPLE AND THE APPRENTICES

(*By* G. Unwin *and* A. Hulme)

COMMUNITY building is the most important aspect of history, but as it is also the most difficult aspect it is the one that has been least studied. This study is at any rate now taken seriously, and it is beginning perhaps to be recognized that those more intensive communities of life, work, and worship which normally tend to absorb the whole personalities of their members are more fundamental to the life and growth of society than the more extensive and attenuated communities such as the nation, the city, and the joint-stock company. To the former class of communities belong all kinship institutions including the modern family, all labour organizations including the modern trade union, all associations for mutual aid, common worship, or other cultural purposes including friendly societies, churches, infant schools, and universities. The individual personality is a focussing point of these communities. Its freedom consists mainly in an open choice of them. And the chief positive significance of that complex body of social changes which we call the Industrial Revolution lies in the more unrestricted formation, the greater mobility, and the wider and more unlimited interaction of these communities.

The workers drawn together in the earliest factories set up in country districts had at first no organized or traditional community life of their own, and were thus thrown into greater dependence on the social initiative of their employer, who if he responded to the call upon his leadership might become in a very real sense a founder of a community. The most remarkable achievement of this kind is that of Robert Owen. He found himself responsible at New Lanark for a population of 1,300, "collected hastily from any place from whence they could be induced to come and the great majority of them idle, intemperate, dishonest, devoid of truth," and for 400 to 500 pauper children who had to be lodged, fed, clothed, and educated. He soon determined that "a reconstruction of the whole establishment would be necessary for his views and for the pecuniary success of the concern." Having overcome all opposition by disinterested devotion to the interests of his workpeople, he became so absorbed in the creative task of community building for its own sake that he abandoned the active management of New Lanark and set out on a world-mission to reconstruct human society as a whole on the same plan.

Samuel Oldknow was not a founder of the same creative order as Robert Owen. He belongs rather to the class represented by the founders of Styal, of Saltaire, and Port Sunlight. He did not, it is true, succeed as these did in making his enterprize pay, but this only makes the extent of the relative permanence of his foundation the more remarkable.

Oldknow's new factory and much of his land lay in the township of Mellor, but the lime-kilns, coal-pit, and the canal when it came to be constructed were on the Cheshire side of the river in Marple, and it was there that the increase of population occasioned by his enterprize mainly took place. The population of Marple, which was only 548 in 1754, had grown to 2,031 by 1801, and it is fairly certain that the greater part of this increase must have dated from the building of Mellor mill in 1790–3. This had also contributed something to the population of Mellor township, which in 1801 numbered 1,670. The two villages thus probably contained in 1790 rather more than two thousand inhabitants. Mellor was much more industrialized than Marple. Radcliffe tells us that

[1] *Life of Robert Owen*, pp. 79, 83, *et seq.*

in 1770 there were only six or seven of the fifty or sixty farmers who did not get their rent partly by spinning and weaving, and the cottagers who were employed entirely in this manner except for a few weeks at harvest must have numbered several hundreds. And before 1790 the "golden age" of the weaver had begun.

Their dwellings [says Radcliffe] and small gardens clean and neat—all the family well clad—the men with each a watch in his pocket, and the women dressed to their own fancy, the church crowded to excess every Sunday—every house well furnished with a clock in elegant mahogany or fancy case—handsome tea-services in Staffordshire ware, with silver or plated sugar tongs and spoons . . . many cottage families had their cow, paying so much for their summer's grass and about a statute acre of land laid out for them in some croft or corner which they dressed up as a meadow for hay in the winter.[1]

In spite of the allowances we have been obliged to make, in view of the recorded facts, for Radcliffe's over-estimate of the prosperity of Mellor at this period, it remains clear that Oldknow was not setting up his factory in an entirely rural district and that he would not find labour there especially abundant or cheap. As far as its positive social effects are concerned the Industrial Revolution may be said to have been almost half accomplished before the Factory System appeared.

Old loom-shops being insufficient [says Radcliffe] every lumber-room, even old barns, cart-houses and outbuildings of any description were repaired, windows broke through the old blank walls and all fitted up for loomshops and this left something in the pockets of every stonemason, carpenter, slater, plasterer, glazier, joiner, &c. as well as the corn-dealer, cheesemonger, butcher, and shopkeepers of every description. The farmers participated as much as any class by the prices they obtained.[2]

[1] W. Radcliffe: *Origin of the New System of Manufacture, etc.*, pp. 59, 67.

[2] Radcliffe, pp. 65–6.

The building of the mill, however, undoubtedly emphasized these conditions. All the surplus of local labour was soon absorbed and a steady inflow of immigrants began. This is shown in a wage list of those engaged in preliminary operations, dated November 28th, 1789, which contains 21 names belonging to the locality and seven names of outsiders. No wage lists for 1790 have survived, but road making and bridge building must have found work for an increasing number.

Of the 204 engaged in May 1791 in erecting the factory, diverting the course of the Goyt, and constructing reservoirs, about half belonged to the locality. Many of the newcomers were lead miners who had been deprived of work by the exhaustion of the mines of the Peak.

By March 1792 the number of workers had increased to 285. Of the additional eighty, about half were drawn from outside sources and the other half consisted of women and girls of the neighbourhood whose labour was beginning to be utilized. In the course of the two years of factory building the names of about six hundred persons appear on the wages lists, but there were never half this number employed at any one time. In the meantime good progress had been made with the Mellor mill.

On the greensward near the ruins at Mellor there lies a large oval stone tablet bearing the initials S. O. and the representation of a weaver's shuttle with the date 1790. This tablet which crowned the work must commemorate the laying of the foundations, as the earliest building recorded is in April 1791. Sixty men had been employed for several days in March in planting trees, and the catching of moles was another preliminary job. As the spring advanced the number of bricklayers engaged increased from fourteen to twenty-five, and a large body of skilled and unskilled labourers were busy throughout the summer months in flagging and sawing, in "walling-in the new river," in getting coals, in threshing oats, and in cutting potatoes. A body of smiths was also at work, and towards the close of the year loads of factory material such as spindle steel were being despatched from Stockport to Mellor.

Most of this information is derived from nine weekly accounts dating from May, August, and September, 1791, and headed "Wages paid at Mellor." The totals vary from £102 to £169, and the wages paid to the workers,

who are nearly all men, ranged from 1/2 to 3/2 a day. There are a number of groups of builders, colliers, etc., whose wages are paid through a contractor or leader. There is no sign as yet of any part of the mill being completed or of any hands at work except women and children picking cotton in their homes.

For the year 1792 there are extant nine similar lists, each covering a fortnight's wages, dating from January, March, August, September, and December, the highest complete total being that of May 31st for £340. These, together with various tradesmen's bills covering the whole year, enable us to follow the progress of the building almost from month to month. In January a dozen plasterers and painters commence operations (at daily wages of 2/– to 2/8) on the ground floor. In May and June the supply of landings, steps, and "risers" shows that the upper stories are being made accessible. In August and September glaziers are using thousands of feet of glass in casements, sashes, and fanlights, and plumbers are laying over a ton of lead piping. Even now there are still men moulding bricks and dressing stone for parts of the structure which are not yet raised, but in other parts of it "flyers" and spindles had been in making, and machines in process of erection, since the end of March. On and after March 3rd the wage list commences with a list of workers, mainly young people, who are obviously not builders but mill-hands, and this nucleus steadily expands into the two-hundred-and-fifty carders, rovers and spinners, reelers, makers-up and winders, smiths and filers who found work at Mellor in 1793. Some rooms of the mill must have been equipped and at work in the preparatory process during the closing months of the year, as actual deliveries of yarn are recorded in December 1792. A separate wage book gives an account of fifty-one outside women cotton pickers at work from June to December. The difficulties involved in organizing this outside labour, in getting the machinery erected and at work, and in conducting the remainder of the building operations are set forth with great clearness in a letter from the manager at Mellor to Oldknow at Stockport.

H. Mason to S. Oldknow :

"Sir, "Mellor, *January* 28, 1793.

"The 3 Bags of Cotton you mention we have rec'd

but I think that what we have Batted here is preferable. The women here are still obstinate & say they will not pick it for 1/6 pr Dozn that is 1½ per lb. H. Furnace has brought in the Cotton she took out on Thursday and she says 1½d. per lb. is too little—however she has taken more out—If you think that we shall continue to bat it in the Building it would be the best to have Fleak made & corded unless you can spare one from Stockport already made—as what we have now is very unfit for the purpose. The reason why rovings have been so scarce is the want of Cotton being picked but I expect that tomorrow we shall have a better supply— The Stretching Frame cannot be altered so as to draw out farther. James Warhurst and his brother will get Timber out for the New Frame as soon as they can get some iron work forwarded but are rather unsettled for want of Iron Work as the Smith who did their work is gone. Steer last week lost 1½ days—Morton lost 2½ days but suppose Morton lost more than he would have done owing to the Death of his Sister.—

"I shall want the price of the Print that Jno brought before I can let any go to any one except the 2 Girls you mentioned. Maycock the Smith says that the Smithy is not ready—the Doors & Windows not being hung—therefore have not been up to the Hall—The Gardener came to work this morning and is now working at the bottom part of the Work which Old Ratcliffe & his company have done. I think he is planting the Beech Hedge. I have been into the Making up Room & told them of their Errors desiring them to be particularly careful for the Future, but think what Twist is to be sent, tomorrow had better not go into the Market as it is not made up in so neat a manner as I could wish, but have desired them to make it up for the future tighter & neater.

"N.B.—I wished to have sent my Cash along by

John but cannot get it finished before he goes however will finish it to Night.

"The Cotton Pickers are all come & are taking out Cotton at 1½d.—tho very reluctantly."

As soon as the mill was built and at work, the social and economic problems involved in the formation of the new community assumed an entirely different form. The demand for adult male labour rapidly decreased. What was now needed was an increased supply of the labour of children and young persons. The unmarried labourers and those with small families who had not found work in the mill would move on to another district where a factory was being built, whilst those with four or five children old enough to work might be tempted to stay, on the chance of a job turning up for themselves. This was the gravest evil threatened by the rise of the factory system. If the father became permanently dependent on the labours of his wife and children, the family, which is the most fundamental of communities, would be demoralized. When we view the matter in retrospect on a large scale the evil seems to have been averted by the pressure of economic forces. In the larger industrial communities of to-day, jobs for men, for women, and for young persons seem to supplement each other automatically by the silent operation of Adam Smith's "invisible hand." But it is a false philosophy which ignores, even under present conditions, the responsibility of human agents for these social adjustments. Under the simpler conditions we are investigating, that responsibility is greater, and the advantage of a detailed record is that it enables us to realize it more clearly.

The total number employed in and about the mill when it was in full work varied from 300 to 350 exclusive of apprentices, and including 15 to 25 mechanics. If we assume 15 of the mechanics to have been adult males and make a liberal allowance for another score of men employed as overlookers in the various departments, we arrive at the conclusion that about one in ten of the workers would be father of a family, and this corresponds fairly well with what we know of similar factories at the same period. The rest of the workers may be divided into family groups of children, young persons and women, numbering from two to six members, but the average is between two and three. Three of these groups out of every four, therefore, had no adult male worker in the factory. The father of the family had to find work outside, and the number of available occupations in Mellor and Marple was limited.

In view of these facts it is clear that the multifarious and ever-extending character of Oldknow's enterprize at Mellor can be partly accounted for on other grounds than megalomania. The thirty to sixty colliers, lime-burners, builders, and farm labourers, etc., for whom he found work on his estate were in many cases the fathers of families employed in his mill. Many of these, no doubt, were only continuing their original occupations, but Oldknow's various undertakings in road making, bridge building, and other improvements enabled him to find employment for a score of additional adult labourers.

A few concrete illustrations of family economy as affected by the conditions we have been attempting to describe may help to make the situation clearer. Samuel Barton was a Marple labourer employed by Oldknow outside the mill in 1793 at 12/– a week. His four children —two girls and two boys—earned 2/6, 3/–, 3/6, and 4/– respectively as spinners and makers-up, and his wife about 2/– a week by picking cotton at home. The total income of the family when in full work was thus 27/– a week. In view of the fact that the allowance fixed by

Quarter Sessions of Oxfordshire in 1795 as absolutely necessary for the support of a married couple and four children, exclusive of rent, was 9/– a week, we may regard the condition of the Barton family as one of relative prosperity as long as the six workers, or even five of them, had regular employment. Similar family groups of five are found earning, in 1793, 25/– and 24/6 and families of four earning 25/– and 19/9. These all contain at least one adult male worker. The four children of Betty Shuttleworth, who is apparently a widow, earn 11/6 a week on an average of six weeks, whilst Betty herself generally makes about 2/6 a week at cotton picking, which produces a family income of 14/– for five workers, with at least six to keep, for there is a younger child at home who occasionally earns a shilling. Another family group of four young persons earn 13/–; one of three, 9/9; and one of two, 6/9. If the fathers of these families, who appear only as drawing the wages of their children, happened to be out of work the condition of the family would be one of want or distress.

The vital factor in the social and economic welfare of the new community was clearly the continuous employment of the fathers of families. The facts hitherto cited afford little light on this point, as they are all taken from the abnormal short period covered by the wage notes in the disastrous year of 1793. It is therefore fortunate that we are enabled by other records to follow the history of many families both of the original inhabitants and of newcomers, and to find evidence that in these cases at least the founder of the community took some pains to keep the heads of families in regular work. One of the most interesting and complete of the account books discovered contains a list of the workers employed on the estate in various capacities for the years 1799-1800, with their wages. The names of 57 men occur in this record belonging to about forty or fifty families which had, in 1794, 151 other members, mainly young persons and women, working in the mill. If we add to these the men still employed in the mill and representing some twenty or thirty families, three-quarters of the 300 would be covered, leaving about twenty heads of families to be accounted for by other employments.

Let us take a single case by way of illustration. A family of immigrants named Derbyshire, possibly numbering ten or twelve but containing eight workers, begins to appear on the wage lists in 1792. The father and eldest son are at first engaged in erecting machinery at 2/– and 1/0 a day. Another young man and three boys work as labourers at rates varying from 6*d.* to 1/6 per day, and some cotton picking is done at home by the mother or one of the two girls. When the mill starts work three of the boys are engaged as carders and one of the girls as a spinner, whilst the other continues to pick cotton at home. From the record of outside workers in 1800 we find that the father and two of the sons are now engaged in constructing lime-kilns, in other building operations, and in gardening, at higher rates than those they had earned in the factory, ranging from 1/10 to 2/10 a day, and one of the girls, now grown up, is also employed in gardening at 1/– a day. Of those working in the mill at this period we have no record.

A dozen similar cases might be cited to show the readiness with which adult male labour was transferred from one job to another, but one will suffice. Ralph Hammond, who began as a cotton spinner, became in turn a coal-miner, a farm labourer, a road-mender, a builder, a gardener, and a woodman. As a rule this change of task made no difference to the rate of pay, which was determined by skill or length of service. James Gibbins, whose work during 1801–3 included the hanging of gates, quarrying, building, potato-getting, road-mending, "soughing," and repairing walls, received the uniform wage of 3/– a day. Of the general rise of wages at this period to meet the increased cost of living illustrations will be given in later chapters.

But whilst Oldknow was striving to some purpose to find work for the fathers of the boys and girls already employed in his mill he was being driven to seek fresh supplies of child labour from sources that would relieve him of that responsibility.

The story of the factory apprentices has been admirably told by Mr. and Mrs. Hammond in one of the chapters of *The Town Labourer*. The exploitation of pauper children for industrial purposes was authorized by that much idealized code of war-time legislation the Statute of Apprentices, and the textile industries, not only of England but of other countries, made considerable use of this expedient long before the rise of the factory system. The invention of spinning machinery that could be tended by children, at a time when the numbers of child paupers were rapidly increasing, led to a more wholesale adoption of this method of employment, which was further encouraged by war conditions and attained an unexampled expansion in the first decade of the 19th century. The evils of the system at its worst were unspeakable. At its best, in the hands of humane and conscientious employers like David Dale, Robert Owen, the Gregs of Styal, and, as we think we may add, Samuel Oldknow, it may perhaps be claimed that the health and morals of the children were better, their happiness greater, and their social outlook brighter in the factory apprentice house than in the workhouse.

The earliest suggestion of parish apprentices at Mellor occurs in May 1791, in an entry in the wage list for the cost of boarding two boys at 4/6 each per week. By

October 1793 the number of apprentices in the factory had grown to 13, and there was no further increase in the next two years. In November 1795, however, we find inquiries being made on Oldknow's behalf in the Parish of Clerkenwell for a supply of children, and on the 24th of that month he was informed that 40 to 50 of both sexes together were available. On December 23rd the friend who was negotiating with the parish authorities writes that they had agreed to furnish the boys and girls, but that some of the children's parents, hearing of the intended destination, had " come crying to beg [to] have their children out again rather than part with them so far off. As there may be obligatory matters to sign on both sides I should think you had better come up to Town yourself to receive them, & if you like me to inquire for more in other Parishes previous to your fetching these, I will do it on your giving me a line." It should be explained that, in London, relief was seldom bestowed without the parish claiming the exclusive right of disposing at their pleasure of all the children of the person receiving relief.[1]

A further letter of January 16th, 1796, informs Oldknow that 35 boys and 35 girls have now been selected and invites him to come and choose as many as he desires, but on his expressing unwillingness to undertake the journey the gentlemen of Clerkenwell agree to send the parish beadle to conduct the children to Mellor and to witness the indentures, as long as Oldknow will meet the expense of returning the children in case any dispute should arise. A doctor had been engaged by the friend who had conducted the negotiations to examine the children and also a few others belonging to the " Liberty of the Rolls " who had been offered on the same terms.

A report of these transactions had evidently soon

[1] Hammond: *The Town Labourer*, p. 145. 1920 Edition.

reached the parish authorities in the neighbourhood of London. On April 27th, 1796, the Overseer of Ashford in Kent wrote on behalf of the inhabitants offering two fine healthy boys, the eldest about ten, the youngest about nine, to be bound till the age of 21, "and as they will not be very willing to part with much money, wish you'd let me know your lowest terms and also whether the children so put out will gain a settlement by such Apprenticeship." At the foot of the letter is a note in Oldknow's handwriting that he would require two guineas to take the boys and also clothes and two shirts. Twelve months later we find him obtaining children from the Rolls Workhouse, and in August 1797 he was in communication with the Committee of the Foundling Hospital about a supply of girls. The only one available at the time was a young woman who was regarded as capable of making the journey alone and was described as "a very good household servant and works very well at her needle." This lends some support to the statement made ten years later, on behalf of the Foundling Hospital Committee, that none of their children "are apprenticed to . . factories except in the case of some girls who proved incorrigible after being put out to service, these if hopeless of amendment are apprenticed to Mr. Oldknow a gentleman in Lancashire, who though he has cotton mills has employed them as domestic servants. Only seven girls have been so apprenticed."

But if Oldknow did not employ the girls of the Foundling Hospital in his factory he had certainly other girl apprentices there. An incomplete list of 24 out of 26, in the year 1807, contains the names of at least eleven girls. In Derbyshire mills the tendency was to employ more girls than boys.[1]

The maximum number of apprentices appears to have

[1] Farey: *Agriculture of Derbyshire*, Vol. III., p. 502.

been reached between 1798 and 1801. An estimate of costs made in the former year includes the maintenance of 100 apprentices at 4/– a week each. In the latter year a teacher was paid for instructing 52 boys, 15 Sundays at 4/4 per Sunday, £3 5*s.* 0*d.* Five other boys were taught for shorter periods. The bill for girls, who were no doubt (as at Holywell) taught separately, has not been preserved. In February 1804, when the next record occurs, the number has declined to 60 and the cost of maintenance increased to 6/– per week. The decline, which would naturally be accounted for by the expiration of terms of service of the large numbers bound ten years earlier, reached a climax at the end of 1806. Between November 22nd and December 6th the number of apprentices decreased from 46 to 26, whilst the number of wage earners increased from 317 to 327, which certainly lends weight to the supposition that many of the old apprentices remained in Oldknow's service after their indentures had expired. That Oldknow did not replenish the dwindling ranks of his apprentices during these years may have been due to exposure of the evils of the system in Parliament and to the proposals put forward for its further restriction. A fresh supply must, however, have been obtained later on, as the number was 43 in January 1812 and 38 in December.

The implication that Oldknow was generally recognized as an exceptionally humane employer is strongly supported by the testimony of Robert Blincoe in a Memoir whose main purpose was to expose the evils of the factory system. Blincoe, who spent some time in Mellor mill during Oldknow's lifetime, states that the apprentices whom he saw at work seemed cheerful and contented, and looked healthy and well, that they were fed with milk porridge and wheaten bread for breakfast, and that all their meals were good and sufficient. The hours worked were from six in the morning till seven in the evening, but this was the shortest working day usual at the time in factories. The worst factories worked from five till eight.

It is possible to learn a good deal about the life these children led during their apprenticeship from the descendants of apprentices who settled at Marple and Mellor.

One old lady of 80, the daughter of an apprentice, gave, in answer to our inquiries,[1] a graphic account of her mother's journey from the Duke of York's Orphanage at Chelsea, in a stage coach, and her life in the 'prentice house. According to her account the children worked long hours, but every day went through exercises in the meadows in front of their house, which kept them in good health. On Sundays they went to church in the morning and evening, by a private road to Marple Church made by Oldknow for this purpose. They were dressed in their best clothes and were accompanied by their employer. Their food was the best that could be procured. They had porridge and bacon for breakfast, meat every day for dinner, puddings or pies on alternate days, and when pigs were killed were regaled with meat pies which were full of meat and had a short crust, such as, the inquirer was solemnly assured, cannot be produced in these days. All the fruit in the orchard was eaten by the children.

In the course of a conversation with one Mellor worthy the evils of the apprenticeship system were mentioned. He listened to the story of the merciless exploitation of children with an air of superior knowledge and then said firmly, "No one ever had owt to complain of at Mellor."

This statement sums up the pride and admiration for

[1] We are indebted for this account to the investigations of Miss F. Collier, M.A.

Mr. Oldknow and all his works which the elderly villagers have inherited from their parents or grandparents. It is certainly a flattering testimony to the character of this factory master who has been in his grave almost 100 years.

But, whatever may be said of the advantages of the factory apprenticeship system at its best as a substitute for the mixed workhouse, it will scarcely be contended that it was a satisfactory substitute for family life, which served as a basis for the better kind of community building described in the earlier portion of this chapter. The cause of this defeat of higher social aims lay not so much in the fact that the cost of maintaining an apprentice was only four shillings a week whilst the average wage of his mill hands worked out at nearly four shillings and eightpence, as in the fact that other enterprizes, not necessarily connected with cotton spinning but arising out of his ambition to found a community, were absorbing more capital than he could pay interest upon. With these enterprizes and their effects upon his fortunes we shall be concerned in the later chapters of this book.

Chapter XII.

OLDKNOW'S PAPER MONEY

ANYONE who is familiar with the office routine of a factory knows how much time and organization have to be devoted to the payment of wages. In a concern with two hundred workpeople thirty years ago this task used to occupy a cashier, an office boy, and half-a-dozen foremen and forewomen during the greater part of the last two days of the week. The mere provision of sufficient small currency was a problem to be solved by special arrangement, and it was not unusual to supplement the agency of the bank by large quantities of copper coinage from a wholesale provision dealer, who had just collected it from the retailers. It is easily conceivable that in the 18th century, when the supply of copper coinage by the Government was sometimes non-existent, and even the supply of silver was frequently inadequate, the problems to be solved weekly or fortnightly by the employer of any large body of wage earners were much greater, quite apart from the more fundamental difficulty of the provision of capital for the purpose. The status of wage-earner was not a new product of the factory system, and Oldknow as a great manufacturer had more workers to provide with regular pay than he had in his later capacity as a master spinner.

A great many of the spinners whom he employed at Anderton and Stockport were indeed small capitalists and as such had to accept payment in two-months bills, though they were often prepared to pay a premium for ready money. But the weavers had to be paid week by week in cash, and as there was no local bank to supply this, even if Oldknow had possessed sufficient credit, it had to be got from anyone, however distant, whose business brought a regular supply of currency and who was prepared to give credit. Oldknow's uncle Thomas was his first banker and sent him at regular intervals sums of £100 or £200 in cash by the carrier concealed in bales of goods. Later on, S. and W. Salte undertook a similar service.

In December 1786 Oldknow established a connection with a firm of bankers at Derby. A few of their letters have survived and show us the provincial banker entering upon the function of providing cash for manufacturers, though with strict limitations.

Thos. Evans & Son, Derby, to S. Oldknow:

"Derby, *December* 4, 1786.

"We have received your favour of the 2nd instant & observe you are supplied with Cash for this week. Our rule is to take common interest for what bills we discount, we never take more nor less and on those terms we can generally choose our customers; which is more satisfactory to us than any trifling advantage we could make in times of necessity. Bills at two months on a good House in London, either your own drawings or endorsements will suit us. We do not much like notes, tho' we have the utmost confidence in the drawer, because if we happen to have any particular occasion for money, they are not readily discounted in London; & indeed our own opinion is against the custom of passing notes. We believe we can supply you with £500 every fortnight, at some times probably we could spare more, which would perhaps make it more an object to you to send for it; as our communication by the Mail is so frequent and quick, we can readily advise you when that is the case.

"We take no half guineas but what draw or are full standards, & Guineas we take none but what draw in general. Sometimes in taking a sum we can not avoid taking two or three standards but never without they are fully such. We apprehend you will have no cause of complaint on that ground as Messrs Arkwrights have had cash from us some years & we never received a complaint nor had a guinea returned. We will reserve for you either £600 or £1000—as you please to direct for the 11th instant, desiring to be favoured with your answer for our government."

January 5, 1787.

"Having notice that some receipts we expected in Cash the next week will be paid in paper, we apprehend that we shall not have any cash to spare before the 23th instant. We therefore have thought it best to give you a line for your government."

An obvious way out of these difficulties was for the banks to issue notes small enough to be used in the payment of wages if they could get the shopkeepers to take them and were themselves prepared to cash them whenever presented. The joint-stock banks of Scotland had led the way in this direction by issuing one-pound notes as early as 1727, though they did not always guarantee payment of cash for them at sight; and in 1763, in the period of speculation that followed the close of the Seven Years' War, the banking companies of Perth, Dundee, and other towns issued quantities of still smaller notes and secured a brief circulation for them by employing agents, who received $\frac{1}{4}$ per cent. commission.[1]

If, however, a large employer of labour had credit enough with the local tradesmen, he might, by issuing his notes to his workpeople, not only solve the problem of providing small currency but economise in the use of his capital. The British Linen Company's Bank grew out of the issue of notes by employers for the payment of weavers, and the Lancashire banking firm of Cunliffe,

[1] Graham: *The One-Pound Note*, p. 81.

Brooks & Co. had a similar origin. Country manufacturers were accustomed to issue notes for sums as low as five shillings.[1]

It will be seen that a remarkable interest attaches to the shop-notes issued in payment of wages at the Mellor mill in 1793–4, from several different points of view. They are symptomatic of an important crisis in the private fortunes of Samuel Oldknow, and constitute an episode in the history of the Truck System. But they have a still wider significance in relation to the history of commercial credit and of paper currency. Enough has been said to show the almost desperate condition of Oldknow's affairs at the beginning of 1793. He had invested an immense capital—for those days—in the fixed forms of land, buildings, and machinery which could not yield any return without the assistance of commercial credit—and owing to the outbreak of war commercial credit had almost ceased for the time being to exist. No fewer than 872 bankruptcies were recorded between November 1792 and July 1793. The problem of credit currency became acute. The country banks, which had multiplied greatly during the previous decade, had produced an over-issue of notes, some of them for such small amounts as to provoke the derisive issue by a Newcastle cobbler of a note for twopence.[2] But the notes even of the sounder banks were now returned on their hands and many were obliged to close their doors. In the chief business centres various plans were adopted for keeping a credit currency afloat and so preventing a complete collapse of commerce and industry. Liverpool with the sanction of Parliament issued municipal notes to the amount of £200,000. A meeting at Manchester resolved that during the emergency

[1] Roth: *The Genesis of Halifax Banking.*

[2] M. Phillips: *A History of Banks, Bankers, and Banking,* pp. 43–60.

it could not be considered disreputable for houses to make payments in their own notes, payable in three months with interest.[1] A committee of business men at Newcastle, after examining the books of the leading bankers, assured the public that their circulation was moderate and their security almost without limit, and a guarantee fund of £320,200 was subscribed by 148 local business men.[2]

One of the leading captains of industry had been beforehand with the business men. Towards the end of December 1792 Lord Chief Justice Kenyon had received a letter from an informer to the effect that Mr. Wilkinson, the famous ironmaster, was paying his workmen at Bersham foundry and in his coal and lead mines every Saturday with assignats. "The Presbyterian tradesmen receive them in payment for goods, by which intercourse they have frequent opportunities to corrupt the principles of that description of men by infusing into their minds the pernicious tenets of Payne's *Rights of Man*." Mr. Wilkinson's sister, adds the informer, married Dr. Priestley. "With what view he circulates the assignats is best known to himself."[3] As the alleged assignat which was enclosed in the letter was made payable at Bersham Furnace and endorsed by Gilbert Gilpin, Mr. Wilkinson's chief clerk, there is every reason for supposing that it was an early sample of the emergency currency called forth by the crisis, and that the Bill for preventing the negotiation of French

[1] Daniels: "The Cotton Trade during the Revolutionary and Napoleonic Wars," in *Transactions of the Manchester Statistical Society*, 1916, 1917.

[2] M. Phillips, *op. cit.*, pp. 48–52.

[3] *Hist. MSS. Comm. Kenyon MSS.* No. 1373. I am indebted for this reference to my colleague, Mr. T. S. Ashton, whose forthcoming work on the ironmasters of the 18th century cites earlier instances of the issue of works currency.

paper money in England which the panic-stricken Government hurried through Parliament at the instance of the Chief Justice was a work of supererogation.

The first of Oldknow's notes that have survived were issued on April 13th, 1793, within a few days of the Manchester meeting and of the Newcastle guarantee. The form taken by these Shop Notes was, to some extent, determined by conditions independent of the commercial crisis. Like other captains of industry who built factories in rural districts, Oldknow had been obliged to organize supplies of the chief necessaries of life for many of his workers. Since the beginning of 1791 he had been providing houses, milk, coals, meat, and beds for an increasing number of transient or permanent employees and deducting the cost from their wages. There is no reason to stigmatize this as "truck." Most, if not all, of the necessaries supplied were produced on Oldknow's own estate and were probably sold more cheaply than they could otherwise have been obtained. Even when he advanced £50, in August 1791, for the purpose of stocking a general shop, it is likely that it was the convenience of the new population and not his own profit that he had chiefly in mind. Enlightened self-interest mingled with higher motives led employers like Robert Owen [1] and the Gregs of Styal to organize supplies without exploiting their workers, and the systems they created passed into the hands of Co-operative Societies.

But however disinterested Oldknow's original scheme of supply may have been, the crisis gave quite a different turn to it. In form it was the application of the principle of payment in kind to the whole of the wages of his employees instead of to a mere fraction of them. In effect it was an appeal to the community at Mellor and Marple, in whose midst he had set up his new enterprize and who

[1] R. Owen: *Autobiography*, p. 87.

might be presumed to have a deep interest in its success, to provide the credit which alone could save it from collapse. In form they were cheques drawn on his own shop for payment in kind at sight. Most of those who received them would, however, not care to take all their wages in that form, but would try to get cash or other goods elsewhere for the balance. In this case the notes became promissory notes of uncertain date, which the workers who received them, or the shopkeepers who honoured or discounted them, must hold till Oldknow was able to provide cash or an equivalent. Sometimes the holder of a bundle of the notes received at once a two months' bill drawn on one of Oldknow's London customers, sometimes he got cash after waiting two months. In this curious way a score of the villagers became involuntary amateur bankers in support of Oldknow's enterprize, though the bulk of the business was undertaken by four or five persons, presumably tradesmen.

What has been hitherto said is, broadly speaking, true of the whole period during which we have any evidence of the notes—from April 1793 to April 1794. But some interesting changes are to be observed within this year both in the form of the notes and in the method of their administration. The first bundle of eighteen are in handwriting throughout, and the following example (apart from the addition sum) represents seventeen of them :

	13	9
1	15	10½
2	9	7½

£0 13 9 "MELLOR, 13 *April*, 1793.

"Pay David Byers or Order Thirteen Shillings and Nine pence in goods to that value.

"Saml. Oldknow.

"At the Shop No. 73."

The one note in exceptional form accounts for the calculation made on the top of the previous note.

£1 15 10½ "MELLOR, 13th *April*, 1793.

"Place to David Byers's Acct One Pound Fifteen Shillings & 10½d for which same Mr. Oldknow will be accountable to you—for whom I am

"yr obed. Servt

"H. Mason."

It is clear from this that David Byers did not wish to spend more at the shop than 13/9 of the 49/7½ that was due to him and asked for a note representing the balance which he could present to another shopkeeper. In the other cases the first form of note was made to serve an equal variety of uses by means of endorsement. For

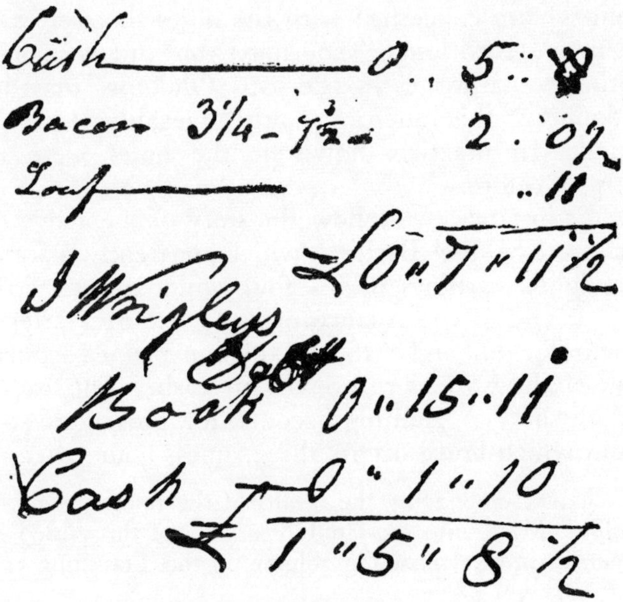

example, Henry Wheeldon, having received a shop-note for £1 5*s*. 8½*d*., expended only 2/11½ in bread and bacon at Oldknow's shop and contrived to get 5/– in cash there. These items were entered on the back of the note and the amount of the balance still due, *i.e.* 17/9, inserted on the face. Wheeldon then went to J. Wrigley's shop and paid off a book debt of 15/11 with the note, receiving the remainder, 1/10, in cash, and these items also were entered on the back of the note (as shown on page 183), which would be subsequently presented by Wrigley to Oldknow for payment of the 17/9.

On May 11th there appears a printed note in quite a different form. (See page 185.)

Over a hundred specimens of notes in this form have been preserved, representing six fortnightly issues during May, June, and July 1793, and facts recorded on their faces and backs are well worthy of careful study.[1] At present we are concerned with the notes themselves. In form they are no longer shop-notes, but orders upon the counting-house for cash. Perhaps Oldknow may have had doubts at this juncture about the legality of payment in kind. In practice, however, the notes were used exactly as before.

If, for instance, we follow the story of the above note made out to Samuel Barton—which apparently represents a fortnight's earnings of his four children—we do not find that any of the balance of 14/4 left over after the payments for rent and coals, and the unexplained deduction of one-sixth, was ever received in cash. The back of the note shows a running account for provisions at the Shop in which bread occurs three times, flour twice, and

[1] A chapter embodying the results of the researches of Miss F. Collier, M.A., into the family economy of this period is to be included in a forthcoming volume on the Transition to the Factory System.

No. [illegible] £0.14.4 Mellor 11 May 1793

At Saml. Oldknows Counting House pay Saml. Barton

or Order the Balance of this Acct.

Dr.	L	S	D
Rent		4	
Cow			
Shambles meat			
Potatoes &c.			
Coals		3	4
Sundry articles			
		4	4
1/6 deduct			
Ballce.		14.4	

Cr.	Days	L	S	D
Mrs Barton			8	–
Mary			7	–
Betty			6	–
Ann			5	–
		1	6	–

H. Mason
for S. O.

meal twice, butter, bacon, cheese, and "print" once. The total is £1 0*s.* 8*d.*, against which the note's balance of 14/4 is received as part payment, and 6/4 remains to be covered by the other note. In other cases, however, payment was partly rendered in cash at The Shop. A great many of the notes have nothing on the back but the endorsement of the person to whom the note was made payable, indicating that he or she has received the value of the balance from some third party. The leading banker in the village appears to have been John Downes, who on July 12th, 1793, presented a bundle of 67 notes bearing dates of May 11th, May 25th, June 8th, and June 22nd, with balances of a total value of £64 18*s.* 9½*d.* and received payment from Oldknow for that amount. John Downes was the man who kept Oldknow's shop, and it seems that he acted at the same time in a private capacity as money-lender to the workpeople. Advances made by him are recorded in the same handwriting as purchases made at the shop.

What evidence, it may be asked, is there as to the extent to which Oldknow made use of these credit notes and of the relief which they afforded in his financial troubles. Fortunately the documents that have survived afford some answer to these questions. Whilst building operations were in full progress during 1791 and 1792 the fortnightly totals of wages and other expenses had averaged between £250 and £300. On March 31st, 1792, they were £340, on September 1st, 1792, £285; but by the latter date a portion of the mill was at work and about £70 of the £285 consisted of wages paid to factory workers and cotton pickers. On April 27th, 1793, when the structure must have been practically completed, though the plumbers and glaziers were still at work and the millwrights were erecting machinery, the fortnightly wages bill came to £184, of which two-thirds was due to mill hands and the remaining £60 to colliers and other outside workers. From these wages £38 12*s.* 8*d.* was deducted for rents and other goods already received by the workers and £129 16*s.* 9½*d.* was paid in shop-notes, leaving about £16 which was paid in some other form. More than nine-tenths of the wages, therefore, were paid, in the first instance, in kind. But in order to meet the case of those who demanded cash the sum of £27 2*s.* 0*d.* was doled out in small instalments which were entered on the back of the shop-notes.

By this time, however, Oldknow, besides accumulating heavy debts of a larger kind, had got into arrears with the small contractors of the neighbourhood, and now that he had got his note system at work the time seemed opportune for meeting their demands. On May 25th, 1793, the account of payments at Mellor rose to £341 10*s.* 5*d.*, of which only about £120 was due to mill hands. Deductions for rents and other supplies already received amounted to £52, notes upon shop to £233, and a bill of exchange for £30 was paid to one of the contractors, leaving about £26 to be accounted for by cash or other modes of payment. In addition to this, however, the holders of shop-notes received small sums in cash amounting to £16 10*s.* 11½*d.* It will be seen that Samuel Oldknow contrived during the crisis of 1793 to carry on his factory by cash payments to his workers amounting to no more than two shillings in the pound.

After July 10th, 1793, the survivals of the balance-sheet form of note cease and the simpler form previously issued comes again into use, and is represented by whole bundles for the first three months of 1794, after which date no more notes of any kind are found. The change in form must have been due to a new system of book-keeping. Some record of the sums owed for rent and other supplies had always been kept on loose sheets, of which many specimens have been preserved. Now that the mill was completed and its office staff organized, the record would naturally be transferred to a ledger and a copy in book form given to each head of a family who received payment in kind, so that a balance sheet did not need to be issued with the note which represented the amount due to the recipient after his debt to the employer had been deducted.

During the last four months of the record we are able to follow the working of the note issue in detail by the help of a book endorsed "Account of Mellor Notes delivered out, beginning the 7th December, 1793." The entries are in the following form :

Del'd	No.	£	s.	d.	When paid	To whom paid
Dec. 7	68		10	3½	14 April	W. Swindells
"	"	1	1	0	13 Dec.	Jno Downes

The number 68 is that of a head of family representing several workers whose wages, after the deductions, come to £1 11*s.* 3½*d.* This sum is paid, no doubt at his own request, in two notes, one for a guinea to be spent or discounted at once, and the other for the remainder. The guinea note is returned to Oldknow within a week through the hands of John Downes, who has already given goods or cash for it, but the other note remains in circulation four months and is presented to Oldknow by W. Swindells, who was a tanner, along with a large bundle of other wage notes accumulated in the interim, probably for the purpose of paying for hides and bark supplied to him out of the produce of the estate. The notes thus issued seem to have covered all payments for wages during the four months of record and were probably applied to other local payments. They were presented for payment by a score of persons. Some of these were workpeople, but more than half the notes returned through the hands of three or four shopkeepers who had honoured them in cash or goods and who were often willing to receive payment from Oldknow in a two months' bill. Perhaps the most significant fact of the whole system is the marked tendency towards the issue of notes for a pound and a guinea which might encourage a wider circulation of them as ordinary currency.

That some of the evils inseparable from the Truck System were experienced at Mellor during this period is shown by the following undated letter :

"MR. Oldknow,

"Sir this is to let you know that I am Extortiant in buying Mr. Downs Shop Goods eavery Article his a penny a pound dearer than I can by at Marpor bridge and a menny of the artles three hapence and 2d worse. Sir I should be greatley obloige to you if youll give me a note go to Marpor bridge it would save me 4s a payment and then I can pay you 4s a payment to wards the ould monny if I ceep trading at Mr Downses I can never pay you for I can hardly Geet vittles

"From your
"Humble Servant
"THOS. AUSTIN."

There is no subject connected with Oldknow's industrial régime on which we could more gladly have further information than that of his paper currency. Unfortunately, not a single shop note survives of later date than 1794. This is the more remarkable in view of the fact that a universal "age of paper" for the nation at large began with the suspension of cash payments in 1797, and lasted till 1821. But probably it is the very universality of paper money at this time that accounts for the absence of any specimens of it in the ruins at Mellor. The notes were now negotiated by the banks and did not return to be honoured directly by Oldknow himself.

We have isolated records of fortnightly wage-payments for the years 1803 and 1804, two continuous records of nearly a year each—in 1805–6 and 1810–11—and these leave us in no doubt of the fact that five-sixths of cash paid in wages at the earlier date consisted of notes of the values of 1/3, 2/6, 5/–, 10/–, 15/–, and 20/–. In 1810–11 the notes were called banknotes and their separate denominations are not stated, so that it is possible that the smaller notes may no longer have been paid. But in 1805-6 the notes which constitute the largest part of cash payments are sometimes called tickets. They are referred to as being "made," "issued," or "renewed" by Oldknow, and banknotes are added as a separate and smaller occasional item.

These facts seem to justify us in assuming that during a considerable part of the Napoleonic War Oldknow paid his workpeople mainly in a currency issued by himself and which we may regard as a further development of the shop notes for £1 which we saw him making use of in 1794. The later notes, presumably, found a wider acceptance and circulation in the neighbourhood than the earlier ones and were not so liable to be discounted, though as part of the general inflation of the period their value, like that of other forms of cash, was tending insensibly to decline.

But there was another feature of earlier conditions which unmistakably survived during the whole twenty years of which we have any record. Payment in kind in the form of deductions from wages for rent, milk, coal, meat, etc., which was almost an inevitable incident in the settlement of the new community, had preceded the emergency notes of 1793–4 and continued after they had developed into a transferable currency. Out of the £250, or thereabouts, earned fortnightly by about three hundred workers at Mellor, some £50 was regularly deducted before payment for rents and other sundries, and as this deduction was not made from all the wages it may in some cases have represented as much as a third of the whole.

This method of payment involved two other survivals —the family or household unit of payment and something corresponding to the balance-sheet wage-note. The evidence for all this does not obtrude itself, but when the wage records are carefully considered it emerges quite conclusively. Each of the fortnightly wage accounts consists of a couple of pages, one of which contains the amounts of wages paid, arranged in four columns without names or other details, and the other indicating in a balance sheet the kinds of currency in which payment as a whole was made and the amount left over. The number of wage payments given on the first page is generally from 150 to 160, or just about half the number of the employees at Mellor, and the amounts of about one-third of the items are too large to represent the fortnightly earnings of individuals. The average fortnightly wage, as the workers consisted largely of young people and women, was about fourteen shillings and sixpence, but at least half the items were over £1, and a great many were over £2 and £3, and sometimes there were several over £4. These items, it is clear, represent the wages of family groups of from two to five persons.

The evidence for the survival of a balance sheet wage-note or of a wage-book in similar form is scarcely less strong. When the total of four columns of wage items has been arrived at, the total of the deductions for rent and other payments in kind is added to it so as to give the total cost of labour for the fortnight. This proves that the deductions have been previously made from the separate items. Now, it would obviously be quite unsatisfactory to offer the sum, say, of £2 11*s.* 8*d.* as representing the various wages of four members of a household minus half a dozen deductions for rent, food, coal, etc. without presenting any particulars of either side of the account.

If we turn to the other page of the wage account we find some evidence of two other survivals closely connected with each other, *i.e.* of the Truck Shop authorized by the employer and of the use of the shopkeeper as a minor source of credit. The total amount of wages to be paid on January 4th, 1806, after deduction of £45 for rent, etc., was £206 12*s.* 1*d.*, and the following is the balance-sheet representing modes and sources of payment :[1]

Dr. Wages.	£	s.	d.	*Cr.*	£	s.	d.
144—15/-	108	0	0	Amt. Wages	206	12	1
128—10/-	64	0	0	6—10/-	3	0	0
27—5/-	6	15	0	6—5/-	1	10	0
46—2/6	5	15	0	21—1/3	1	6	3
46—1/3	2	17	6	Silver	1	2	6
" Mrs. W. "	8	19	2	Copper		8	0
Silver	4	12	6				
Copper	2	0	0				
A. Hibbert		9	1				
Unpaid		10	1				
Notes Issued	10	0	0				
	213	18	4		213	18	10

It may be remembered that the provision shop at which many of the notes of 1793 were honoured was kept by Mrs. Wrigley, and as the name of " Mrs. W." appears

[1] The totals given above are as they appear in the original.

regularly throughout the period as indicating a source out of which wages are paid, we may assume that the broken amounts with which she is credited represent a number of notes for broken amounts drawn upon her and paid instead of cash to those in her debt.

As clear proof of the banking functions performed by the village shopkeeper we have a little book containing a complete account from 1801 to 1811 of Joshua Wrigley, junior, the real proprietor of Mrs. Wrigley's shop, with Samuel Oldknow & Co. The transactions recorded in it are entirely advances of cash by Wrigley to Oldknow, mainly on the fortnightly wage days, and the repayment on the same day, or soon after, by a two months' bill, with the addition of interest at 5 per cent. The sum thus advanced represents the money taken in the shop during the previous fortnight, to which, however, there is added the amount of the credit notes made out to "Mrs. W." as appearing in the wage account. On January 4th, 1806, for example, Joshua Wrigley is credited in his account book with £49 19*s.* 2*d.*, which we may take to mean that in addition to the amount £8 19*s.* 2*d.* of the notes made out to "Mrs. W." and paid to her customers, he has advanced £41 in various forms of currency for the payment of wages. In the wage accounts of 1810 similar advances are made by other local shopkeepers bearing the same names as Oldknow's creditors of 1794.

Chapter XIII.

MELLOR MILL AT WORK

ALTHOUGH by the middle of 1795 Oldknow had given up his manufacture of muslins, had sold his Heaton Mersey works, and let his Stockport premises for £600 a year, his investments at Mellor, where he had now taken up his residence, must have soon more than made up for the mobilization of his capital. As one of the leading promoters of the Peak Forest Canal he acquired 52 shares of £100 each and later increased his holding to 216 shares ; his lime works involved an outlay of £4,500, and he must have spent much larger sums on building and other improvements.

The hope expressed by Arkwright in April 1795, when acknowledging the receipt of interest and of a new bond for £11,500, that Oldknow is about to convert something into money and will lessen this vast debt very soon, is belied by the record on August 3rd, 1796, of a new mortgage of the Mellor properties for £32,000 to Arkwright, who has apparently taken over some of Oldknow's other debts and retains his position as leading creditor. If all this expenditure is regarded as arising out of the erection of a spinning-mill which contained so far only 8,464 spindles it looks decidedly like a case of over-capitalization. And this leads us to consider some of Oldknow's own estimates of the cost of running his mill and of its monthly output, which exist from 1797 onwards in considerable numbers and possess perhaps a unique value as statistics of the earliest recorded period of the factory system

The two documents placed together for comparison on the opposite page are the earliest and perhaps the most striking examples of this class of records The first is a statistical statement accompanied by an estimate referring to January 1797. The second is written on the back of

MELLOR MILL MONTHLY STATEMENT FROM JANY. 1ST TO 31ST, 1797

No.	Spindles Set up	Total Spindles	Average Spindles Running	Cotton to the Cards	Waste to the Cards	Hks. Spun	Lbs. Spun	Wages Paid		
								£	s.	d.
1	2608									
2	2464	8464	6000	15842–2	4452	340380	17060	405	2	10
3	2816									
4	576									

CALCULATION ON THE ABOVE ACCOUNT

		£	s.	d.
Cotton-wool to the Cards,	15,842 lbs. 2 ozs. at 2/1½	1683	1	9
Waste to the Cards,	4,452 lbs. 0 ozs. at 1/4	296	16	0
	20,294 lbs. 2 ozs. ..	1979	17	9
340,380 Hks., 17,060 lbs Spun—Contingent Expences 1/- per lb.		853	0	0
	Balance or apparent Gain	11	16	5½
		£2844	14	2½
8,450 lbs. Spun Yarn, 4/7 lb.		1943	18	2
	Deduct 10 per cent.	194	7	9¾
		1749	10	4¼
8,610 lbs. Spun Yarn, Present Money		992	5	5
17,060 lbs.		2741	15	9¼
3,234 lbs. Waste at various prices		102	18	5
20,294 lbs.		£2844	14	2¼

The Contingt. Expences are taken separately as below:	d.
Hands in the mill, per lb. ..	5½
Management, Clerk, Smiths, Joiners, etc.	1¼
Wear and Tear	1¾
Interest of Money	2
Rent, Insurance, Carriage of Stores, Postage, etc. ..	1½
Per lb.	12

The 10 per cent. Deducted:	per cent.
Discounts to purchasers ..	5
Commission for selling ..	2½
Risque of Debts and Carriage to Manchester	2½
Per cent.	10

ROUGH ESTIMATE OF COSTS ON BACK OF LETTER DATED AUGUST 24TH, 1798

Rent of Mellor Mill 1,600 (£) a year

	£	s.	d.
Rent of a week is 2*d*. a pound on 3,700 lb.	30	16	8
Rent or interest of machinery is, at 10 per cent. on £16,000, the same as the rent, and on 3,700 lb. is 2*d*. a pound	30	16	8
Management, Clerks, Smiths, Joiners, etc., Insurance, Carriage, Postage, taken at 2*d*. a lb. on 3,700 spun per week	30	16	8
Per week or 6*d*. a lb. on 3,700	92	10	0
Wages per week:			
318 mill hands	74	0	0
100 apprentices	20	0	0
	94	0	0

That 12*d*. a lb. on 3,700 is the real cost of Spinning and £1 10*s*. 0*d*. over.

a letter dated August 24th, 1798, and is merely a rough estimate of costs, but it supplies certain facts not given in the earlier statement. Let us first examine the statistics and then compare the estimates.

The first point to strike us is that a mill so celebrated in its day should have so small a number of spindles, and that over a quarter of those set up were idle. It will be noticed that one of the four rooms is only partially equipped with machinery. The average count of yarn spun, as shown by the division of hanks into lbs., is only just over "twenties." But the most vital fact is the smallness of the "balance or apparent gain," which is only £11 16*s.* 5½*d.* for the month, or less than £3 a week. The explanation apparently lies in the statement that whilst half the yarn, which is presumably already sold, is subject only to a discount of 10 per cent., the other half, which is presumably unsold, is valued at only 50 per cent. of its listed price.

If we turn to the two estimates of costs, the first of which is in the writing of Oldknow's cashier and the second roughly drafted by himself, several important discrepancies may be noticed. The amount of output presupposed is almost the same in both cases—17,060lbs., spun in a month of four and a half weeks, is about 3,790 a week. The wages item in the second estimate accounts for rather more than 6*d.* in the cost of each pound of yarn, as against 5½*d.* in the first estimate. But the really important difference between the two calculations is in respect of the amounts for rent and for wear and tear. In the first statement, rent, along with insurance, carriage of stores, postage, etc., is estimated to cost 1½*d.* per lb. If we subtract ¾*d.* for the other items, which seems to be the amount allowed for them in the second estimate, only ¾*d.* per lb. is left to cover the rent for which 2*d.* per lb. is allowed in the same document. In other words, the estimate for rent is suddenly raised from £600 a year to £1,600 a year; whilst at the same time an estimate of £1,400 a year for wear and tear is dropped out altogether.

Without presuming to give an expert judgment, we may perhaps venture the opinion that the statement of January 1797 indicates a very unsatisfactory condition of affairs, and that the rough estimate drafted by Oldknow in August 1798 looks like the effort of a sanguine temperament to put a more cheerful if a somewhat subjective interpretation upon the facts. The essential point of the estimate is that he is trying to make the spinning business account for more of the capital sunk in the Mellor estate than it can possibly bear.

The times were bad, however, even for those who had not invested too much in real property. The worst crisis of the war, which led to the suspension of cash payments in February 1797, caused a complete stagnation of business which lasted well into 1798. Muslins as well as yarn were being offered at 40 per cent. below cost for cash.

A letter from Arkwright in February 1798 faithfully reflects this condition of affairs and casts further light on Oldknow's troubles.

"No favourable circumstances have arisen in business to induce me to alter the sentiments I expressed when I last wrote to you, indeed everything seems to conspire to extinguish the little prospect there then was of my taking up the idea you have been suggesting. It would be downright madness in me to engage myself—I dread the consequences that seem likely to follow the present situation of trade. Many mills are giving over. In this neighbourhood Mr Strutt's have shortened their time of working, and so has Mr Nightingale—I have been so much undercut that I have made little sales for this long time. Nobody adheres to their lists and what steps to

WHEREAS

The horrid and impious Vice of profane CURSING and SWEARING,---and the Habits of Losing Time,---and DRUNKENNESS,---are become so frequent and notorious; that unless speedily checked, they may justly provoke the Divine Vengeance to increase the Calamities these Nations now labour under.

NOTICE is hereby given,

That all the Hands in the Service of

SAMUEL OLDKNOW

working in his Mill, or elsewhere, must be subject to the following RULE:

That when any person, either Man, Woman or Child, is heard to CURSE or SWEAR, the same shall forfeit One Shilling.---And when any Hand is absent from Work, (unless unavoidably detained by Sickness, or Leave being first obtained), the same shall forfeit as many Hours of Work as have been lost; and if by the Job or Piece, after the Rate of 2s. 6d. per Day.---Such Forfeitures to be put into a Box, and distributed to the Sick and Necessitous, at the discretion of their Employer.

MELLOR, 1st December, 1797.

CLARKE, PRINTER, STOCKPORT.

A NOTICE TO WORKPEOPLE

take I do not know . . . I am sorry it is not convenient to you to remit me the interest . . . It will accumulate if you do not do it, and that is what it should not do. I shall at all times be glad to see you here, when it is convenient to you, but upon the business you allude to, it is unnecessary, for I could not say anything new on the subject. I have a list of spinners who sell at a great disc't, and you are amongst them : indeed you cannot avoid doing so, [money] you must have—It seems to me that trade cannot mend, till there is a peace, or till one half or two-thirds or perhaps three-fourths of the mills have given over working."

Upon what basis of co-operation Arkwright's aid was sought there is no record, but as Oldknow still retained possession of his Stockport properties it is probable that he had not abandoned hope of reviving the wider plans of 1792. To Arkwright's refusal must certainly be mainly attributed the offer in 1798 of those properties and others in Marple and Mellor for sale by auction, the printed notice of which furnishes the most authentic information we possess about the premises in which Oldknow had established the factory system at Stockport. In 1784 he had paid £550 for the house in which he set up his business and the land on which it stood. The rent of the property was greater in 1798 than its capital value had been in 1784. More land had probably been acquired, but the chief increase in value was due to investment in buildings and machinery.

Lot 1.—All those extensive PREMISES, late in the occupation of Mr Oldknow and now occupied by Messrs. Parker, Sykes and Co., pleasantly situated in the Higher Hillgate, Stockport; consisting of a good House, Stabling, Offices, Garden, and commodious Buildings five stories high, now used for Spinning, and the Manufacture of Muslins and other Piece

Goods and has every necessary Convenience for making One Thousand Pieces per week; the site of which contains about 7640 square yards, part freehold and part lease-hold for long terms of years. Together with an excellent STEAM ENGINE of Messrs Boulton and Watt's constructing; and many valuable Fixtures which will be sold therewith.

Lot 2.—Five good, substantial BRICK DWELLINGS, with a Picking-room and Loom-House standing on lands held for ninety-nine years under the Rector of Stockport, adjoining the above.

The first of these lots is let for £520 a year and the second for £80; making together a neat rent of £600 a year.

Some sales were effected, but the Stockport premises were presumably withdrawn because the reserve price was not reached, as they were not finally disposed of till 1801, when they became the scene of William Radcliffe's famous inauguration of the dressing machine. In the meantime the most decisive of the many crises in Oldknow's career had been reached. During the years 1798–1800 we hear little about the spinning-mill, and a great deal about the improvement of the Mellor estate. High-farming, road-making, bridge-building, and canal-construction were now in full swing. The wage list of Oldknow's outside workers doubled, his canal shares multiplied by four. On none of these investments was he likely to get a rapid return. It is not surprising, therefore, to find that in September 1800, whilst the value of his estate was estimated at £83,794 10*s.* 8*d.*, the amount of his debts was £97,400 18*s.* 5*d.*, so that he was in fact insolvent.

From an open declaration of bankruptcy he was saved by the intervention of Arkwright, who now entered into partnership with him on terms which, whilst leaving ostensible possession and actual management to Oldknow, transferred the essentials of ownership to the other partner, who had in fact advanced most of the capital. The right to receive the interest and principal of £32,000 previously lent to Oldknow and covered by mortgages was handed over to a third party to be held in trust for Arkwright, who was to advance at 5 per cent. another £40,000, making altogether seven-eighths of the value of the estate, for the purpose of paying off debts and restarting the business. Over and above the 5 per cent. on his capital he was to be entitled to half the profits, and to have the right of intervening, if he so desired, without the obligation of taking any active part in the management. Books were to be kept and yearly accounts rendered in such a way as to make a regular inspection easy. Oldknow on his part was to live rent free at Mellor and give his whole time to the management. He was not to draw more than £400, or one tenth of the clear gain, for his yearly expenses, but to put the rest of his share of the profits into the capital account. He was also to be accountable to the partnership for the difference between the rents of such parts of the estate as were let to tenants and the amount of interest payable to mortgagees, "and having some time theretofore entered into contracts for the sale of cotton twist at a lower price than what the same kind of article was then or might thereafter be selling for," he was to make good whatever loss might arise to the partnership in consequence of such contracts. The conditions of the partnership, which was to last fourteen years unless dissolved earlier by common consent, are such as might have been prompted by real consideration on the part of Arkwright for an old friend who was an almost incorrigibly bad man of business. They prevented the break up of the estate and the collapse of all Oldknow's deeply cherished projects, secured to him the actual management and ostensible possession, and held out the prospect of a resumption of full ownership if the debt which had been accumulated could be cleared off.

That prospect, indeed, was never realized. Perhaps it is too much to expect that a man of sanguine temperament who has acquired habits of profusion should become a strict economist after reaching the age of forty-five. When the partnership was dissolved, on January 1st, 1805, the estimated value of the estate had nearly doubled and was £160,584 but the debts had also risen to £156,333, of which £143,000 was owing to Arkwright. Oldknow was thus left in a condition of solvency but of heavy indebtedness, the amount of which continued to increase. He remained in possession of the estate with all the prestige of a county magnate who was at the same time the founder of an industrial community, till his death, when the ownership naturally passed to his chief creditor and former partner by whose capital it had been mainly acquired.

The only important evidence furnished by the records of the new conditions introduced by the partnership and not affected by its dissolution is the greater clearness of the accounts. Several series of monthly sheets of statistics were drawn up obviously for the information of Arkwright, and copies of many of these, belonging to the years 1804–7, 1811–12, together with a part of a ledger for 1800–2 and large fragments of a wage-record for 1805–6 and 1810–11, combine to give a fairly full account of the quantity and quality of output, the cost of production, the names of customers, the number and average wages of the workers, and other details of the new factory economy.

The main facts about the output are soon told. The spindles at work increased from 6,000 in 1797 to 10,080 probably before 1804, as in March of that year there is the maximum record of labour—432 hands in the mill, including 60 apprentices, besides (judging by the wages paid) about a score of mechanics and a hundred pickers, making a total of about 550 workers. There was a small but fairly continuous decline in these numbers, and at the end of 1812 the hands in the mill were 345, including 38 apprentices, the mechanics 15, and the pickers only about a score, as their wages had dropped from £83 a month to £14. The output measured in weight remained almost exactly the same in 1812 as in 1797, *i.e.* 17,066lbs. of yarn in a month in place of 17,060; but as measured in fineness of count it had nearly doubled, being 627,459 hanks instead of 340,380, or an average of 36¾ to the lb., instead of 20. The price obtained for the higher counts was considerably less than that obtained for the lower counts fifteen years before, though the labour cost per hand per week had risen from 4/6 to 7/4 in accordance with a general rise of prices.

A more intimate sense of contact with the working of the business is derived from a series of three little memorandum books dated 1803, 1805, and 1807, which must have been carried in Oldknow's pocket to Manchester, and which contain lists of debts owed by or to himself and of bills falling due, notes of orders received, contracts made, and such lesser matters as "Wanted for Man'r Canal Warehouse, 4 Strong hooks suitable for fixing in stone to hang 2 stable doors on." These books serve to illustrate his methods of business. They show that he dealt largely in warps as well as twist, and indicate the names of his leading customers for both, amongst whom are H. & W. Fielden, and Birley & Hornby of Blackburn, and Radcliffe & Ross of Stockport.

Chapter XIV.

HIGH-FARMING AT MELLOR

(*By* A. Hulme)

BOTTOMS Hall, Oldknow's principal farm, which also served as an apprentice house, stands at the junction of two roads, one leading up to Mellor village and the other to Linnen Clough farm, Tarden, and Cobden Edge. The precise date of this building we are unable to give. All we can say is that the style of architecture and the substantial workmanship stamp it indisputably as Oldknow's work. The outbuildings for the accommodation of cattle are a revelation of forethought and enterprize. The licking troughs in the shippons are hewn out of solid stone. There is also a diminutive shippon for the accommodation of calves. By means of a boiler and pipes laid to several stone containers—which still exist, but are now never used—Oldknow was enabled to steam the cattle food when so desired. The stream from the Linnen Clough reservoir, which runs past this farm, worked a small water-wheel connected with a shaft running through the farm buildings. This was utilized for cutting chop (hay), turnips, etc., and possibly for churning. The ruins of the wheel and the shaft are still to be seen.

Early in 1793 a company of men were engaged on the Windybottom portion of his estate in cutting a new road and using the soil in levelling up other parts of this land. At the same time the Goyt Cliff Torr estate on the opposite side of the river was drained and stones were cleared off the meadows.

The discovery of several of his seed bills for the spring of 1793 indicates the variety of vegetables grown, of which the following is a selection : Turkey long pod beans, marrow peas, speckled kidney peas, Cos lettuce, cabbage lettuce, long prickley cucumber, green Turkey cucumber, sugar loaf cabbage, Italian brocoli, Welsh onion, Salmon radish, early Dutch turnip, new Strasburg onion, orange carrot, white beet, etc.

A broadsheet discovered in the Mellor mill ruins and issued in 1793 by the "Society of Agriculture at Manchester"—which had amongst its subscribers Robert Peel, James Watt, Samuel Oldknow, Arthur Young, and others—offers prizes for planting the greatest number of poplars, hoeing the greatest number of turnips, bringing up the greatest number of legitimate children on a rental of not more than £4 a year, for discovering the best methods of destroying rats and cockchafers, preserving cabbages, and keeping rivers within bounds.

Oldknow provided facilities for a number of his workpeople and tenants to keep cattle. He also kept a herd of milking cows to supply the needs of his factory workers and deducted the amount of their purchases from their wages. His particular fancy was a cross between his Devon bull and Shorthorn cows, as he considered this better adapted for milk producing than either of the two breeds mentioned. For the purpose of supplying the factory hands and tenants with meat, bullocks were reared, and some of these were used for draught, a practice then dying out. A reference to this appears in the 1798 memorandum book: "The Oxen are to continue to remove the heaps of Sweepings from the road Side."

Another document records the cost of maintaining a sadler—1/– per day—engaged in mending "the gearing at the Windebotam and Chapel houses for the Bullocks and Horses." It may be that the oxen were used to avoid the horse tax.

In sheep farming Oldknow showed a live interest. In addition to the native sheep common to the district known as "Woodland" or Moorland sheep he had a number of Sherwood Forest ewes, which produced very fine wool and were considered superior to the local variety. He also kept rams of Bakewell's "New Leicester" breed, and one of these gained a cup at the Manchester Agricultural Society in October 1808. Another feature of his flock was a number of "South Down" ewes acquired from Francis Eyre of Hassop in 1809. These he crossed with a Spanish breed known as "Merino" purchased at King George III.'s sale, and at Colonel Downie's Liverpool sale in September 1810. He was the earliest to introduce the "Merino" breed into the Peak district and his example was quickly followed by many other sheep-rearers in Derbyshire.

In Oldknow's day the common practice of the wool-staplers in Derbyshire was to go round to certain farmers and sheep rearers and bargain for the wool. Also farmers called on the wool-staplers at their houses in secret. This method of sale, coupled with the fact that no wool fairs existed in the county, brought about considerable differences in prices, as there was no competition in the open market.[1] Thus it was that in 1808–9, whilst some sheep-rearers only obtained 1/6 a lb., Oldknow bargained for 3/6 per lb. with the wool-stapler, washed on the sheep's backs.[2] Without doubt Oldknow's wool production helped to supply the neighbouring woollen cloth factories, of which there were a number situated at Glossop, Hayfield, Ludworth, and Phoside, both for yarn spinning and weaving. Also hat-making, mainly a domestic industry in the surrounding towns and villages, absorbed its quota of the local wool production, as there was a considerable demand for lambs' wool for this purpose.

The opening of the lime kilns in 1797 was not only an important event in Oldknow's career as an agricul-

[1] Farey: *Minerals and Agriculture of Derbyshire*, Vol. III., p. 137. [2] *Ibid.*, Vol. III., p. 126.

turist but had a far-reaching effect in stimulating and improving agriculture in a much wider sphere than Marple and Mellor, as farmers from the surrounding towns and villages purchased large quantities of lime from the Marple kilns. Oldknow's own career as an experimental and scientific agriculturist received a considerable impulse from the opening of these lime kilns. Previous to this date references to lime for agricultural purposes are few. Oldknow's land on the lower slopes of the valley was well wooded, many references to the felling of large numbers of trees occurring in the records from time to time. Owing to the large quantities of lime used on this land we conclude that it must have been rich in decayed vegetation, as lime applied to poor, light, or sandy soil would only accentuate its povertv. Of the numerous references to the spreading of lime on different portions of his estate the most interesting is a note which appears in a memorandum book for 1798 :

Work to be done by the Lime Horses :

1st. To finish Limeing the fallow.
2nd. To set one Butt over with Lime Ashes and another with Lime in the Rushy field by way of Experiment to see if either are useful and if so which answers the best.
3rd. To Carry Lime into the Brows for mixing up with the guttering Clods.
4th. To Lime over Two Pasture Fields for Miss Shaw and finish the bench Hay.

In the spring of 1798 a considerable amount of " Gutter Ploughing "[1] was in progress on the Bottoms Hall, Linnen Clough, and Goyt Cliff Torr estates, the clods in every instance being mixed with lime and heaped up to be used as manure when required. At the same

[1] Surface draining on clay land.

time several fields adjacent to Linnen Clough farm, then known as the Cobden Meadow, the Rushy Field, and the Top Festerstocks, were ploughed and sown with oats, and fields on the Collier Fold estate above the canal were drained and made ready for " limeing." A note in the 1798 memorandum book states that Ralph Hammond and Richard Griffis are to " regulate the piece of ground, make good the Fence, and take off all the water, so as to prepare for Limeing the Piece of Ground above the Railway." This refers to the inclined plane railway connecting the summit level of the canal with the bottom level at the aqueduct.

At this time, evidently a slack period, a number of mill hands had been working intermittently in making and levelling land at the back of Oldknow's residence, " Mellor Lodge," a piece of ground in the orchard was also sloped and planted with potatoes, and fresh-made land at the back of the factory was sown with clover and hay seed.

Wheat was not usually grown in the Mellor district, although Oldknow appears to have cultivated this cereal and also to have grown a considerable quantity of oats. In cultivating wheat his first operation was to lime the land, usually in September, and to plough it in immediately, probably to avoid the wet weather, as wet lime had a tendency to burn the horses' feet. The threshing of the grain was performed by the flail, a laborious but effective implement. In 1798 Oldknow paid William Armfield 6*d*. a measure[1] for " threshing the grain at or in Linnen Clough barn " ; then follows a footnote, " He is likewise to thresh the grain that is for the fowles and take care to deliver the same at the mill free of any expence to

[1] A measure in Derbyshire was anything between 32 and 36 quarts. It was usual to provide the thresher with small beer in addition to his pay.

Mr Oldknow." There is nothing recorded to suggest that threshing machines [1] were ever employed, although these machines had been introduced into Derbyshire by a Mr. John Lingard, of Wormhill, as early as 1797.[2]

From the Manchester Police Commissioners he obtained large quantities of manure, which was conveyed to Marple by canal. From the privies of his factory and the kitchen of his apprentice house he laid drains to a large cesspool or well, and by means of a chain pump raised the manure water into landers and water-carts, which distributed it on to his grass lands. The Board of Agriculture report on Derbyshire for 1813 states that in the whole of the county the only instance noticed of raising water by machinery for irrigation purposes was at Mr. Oldknow's Mellor mill.

Conveniently situated on Oldknow's estate were two valuable quarries of "freestone," so called because the stone was capable of being hewn with almost equal ease in any direction. On account of its dressing qualities, its use as building stone was much appreciated, and Oldknow, with his usual foresight, made the most of it. Of these two the Goyt Cliff Torr quarry, on the Cheshire side of the river, provided building material for his Marple estate [3] and the other, the Cobden quarry on the Derbyshire side, appears to have been used mainly for the Mellor improvements. Following the building of the factory, Stone row, Brick row, and Mellor Lodge—all of which were finished by the end of 1794—building operations were apparently confined to the Lime kilns. In 1800, however, activities again commenced, and in wages alone —apart from the cost of materials—Oldknow spent £400

[1] Invented by Andrew Meikle about 1786.

[2] Farey: *Agriculture of Derbyshire*, Vol. II., p. 49.

[3] Marple Church and Tower were built with stone from this quarry.

during this year in buildings and repairs. £176 of this sum was absorbed by the wages of the twenty masons building the bridge which carries the road from Bottoms Hall over the small stream at the head of the Mill-dam, to the road leading up to Mellor. This was accomplished during the autumn of the year. Amongst the remaining items are improvements to Mellor Lodge, the Chapel Houses, and the Navigation Inn. There is also an entry for building a "Soup Shop."

During 1801, when new lime kilns were being erected, the following buildings were also commenced: Garden house, Marple Lodge, cottages known as "Canal Buildings," behind the Navigation Inn, and the house at the Flood Gates. At the same time many wages slips for quarrying stone from the previously mentioned quarries and records of sawing up timber confirm the above statements. In building his factory and cottages and in fencing and effecting these further improvements Oldknow had been obliged to fell large quantities of timber, in the place of which he set up new plantations. In carrying out this branch of rural economy he generally planted the following kinds of trees: Abele, Alder, Beech, Elm (English and Wych), Larch, Mountain Ash, Oak, Poplar (Balsam and Black Italian), Scotch Fir, Spanish Chestnut, and Sycamore. Instead of trusting to memory, haphazard ways, and other folk's opinions, he adopted systematic methods, by numbering and keeping separate accounts of a large number of young trees growing in these plantations, and in this way obtained scientific and reliable data as to the effect of pruning and thinning, etc., upon their growth and progress. Larch, on account of its speedy growth, its recognized value as fencing timber, and its ready sale, had become Oldknow's particular care. There were plantations of larch on the road from New Marple to Longhouse Lane.

In connection with his scheme for afforestation a

nursery was established for the raising of young plants and trees. The earliest reference to this is in February 1803, when five men are employed building a wall round the nursery. The next year Miss Shaw purchased at the undermentioned prices :

700 Quickset . . .	13	0
200 Spruce	5	0
400 Larch	8	0
200 Oak	6	0
300 Elms	12	0

A further note in January 1804 refers to forest trees valued at £20 being taken from the garden. The only other surviving record of the sale of young trees from the nursery appears in the garden book for March 1811, as follows :

Poplars taken out of the Garden . .	18	13	6
Firs taken out of the garden . . .	8	0	0

Though we possess no complete account of the sale of timber[1] Oldknow must have derived a fairly large

[1] In 1811 Oldknow paid part of his Mellor taxes amounting to £45 18*s*. 1½*d*. in timber.

Major Oldknow Dr. I Immanuel Wild, Collector.

Window Duty	8	15	0
House Tax		8	0
Male Servants	3	19	6
Carriages	5	0	0
Horses for Riding or Drawing Carriages	4	0	0
Other Horses & Mules	11	17	6
Dogs		13	0
Armorial Bearings	1	1	0
Ten per cent. on the above	3	11	5¾
Land Tax	4	12	8
Expenses of Case	2	0	0
	45	18	1¾
Allowance for Children			
Received in *Timber*			

revenue from this source, for in September 1805 he records £94 4*s.* 9*d.* from the " sale of timber " and during February 1806 £1,708 10*s.* 5*d.*, and November of the same year £969 5*s.* 6*d.*

In this district the timber was usually sold standing so as to avoid the auction duty. The more common method was to sell " by ticket." The buyers and the vendor being assembled at a public house, the vendor put a folded ticket on which was written his price of the lot about to be sold into a glass on the table, each of the buyers did the same, and then the vendor opened all the tickets but his own and declared the name of the highest bidder. A second delivery of tickets by the buyers then took place, and the name of the highest bidder amongst them was again declared, and then a third delivery, which according to the practice decided the sale, unless on opening the vendor's ticket none of the biddings came up to it, when the sale was declared void. The seller could of course accept the highest bid, even if it did not reach the price of his ticket, if he so desired.

The sale of alder bark, from which was obtained a very good red dye, was still another fruitful source of income, the Manchester dyers paying from £5 to £6 10*s.* 0*d.* a ton for this useful commodity. The alder poles were then sold at prices ranging from 1/– to 1/4 a foot, to be used for turning bobbins and spindles.

A number of tan-yards in the villages of Chapel-en-le-Frith, Charlesworth, and Hayfield, and the hamlet of Whittle, obtained supplies of oak bark from Oldknow. A quantity was also bought by a local tanner, William Swindells, who purchased the hides and skins left after Oldknow had supplied his workpeople and apprentices with meat. The effects of the Napoleonic War on prices are strikingly illustrated in this particular case. The usual price of oak bark in 1792 was £5 per ton. By 1802 the price had gradually risen to £7 10*s.* 0*d.*, and in 1809 Oldknow realized £15 on each ton delivered to the tanners,[1] having paid for the peeling at 25/– a ton. He very wisely employed his own peelers to ensure careful handling of the timber, and consequently less damage was done than if he had allowed the tanners to employ their own men, as was often the custom. A further string to his bow was the planting of osiers on several islands in the Mellor mill-dam, which were cut annually and sold at a good profit. These trees also served a useful purpose in harbouring wild ducks in considerable numbers. The records show that osiers were also cultivated by Oldknow at Combs reservoir, and sold to the local basket and whisket makers, an industry then carried on in Chapel-en-le-Frith.

A rich sheltered portion of land three acres in extent, lying between the river and the foot of the Goyt Cliff below the road, provided an ideal situation for a garden, and here Oldknow built his garden house and employed a professional gardener, who cultivated, gathered, and delivered all the useful vegetables and garden fruits in season to the workpeople and tenants, and rendered an account once a fortnight to the factory manager, who deducted the amounts from the wages of the various purchasers. Attached to the garden house were proper rooms for drying, cleaning, and preserving fruits and seeds, and also a wool drying chamber for use in the sheep shearing season.

The total receipts from garden produce for the years 1804, 1809, and 1810 are as follows:

1804	1809	1810
£	£	£
262	285	329

[1] Farey: *Agriculture of Derbyshire*, Vol. II., p. 334.

Generally speaking, the local inhabitants and work-people consumed from one-half to two-thirds, and the apprentices roughly one-sixth, of the total production The residue was partly used by Oldknow's own household, the cattle, sheep, and pigs accounting for the remainder.

In planting fruit trees by the side of the factory and training them up the walls Oldknow not only provided them with support and shelter, but also added considerably to the appearance of the building.

Oldknow's experiments in agriculture were so successful as to merit special mention by the writer of the Board of Agriculture report on Derbyshire (1813), who referred to Oldknow's delightful grounds as the best example of well-conducted improvements with which he was anywhere acquainted. A biographical notice in the *Gentleman's Magazine* of November 1828 claims that as an agriculturist he was surpassed by none for care and judgment in the selection of stock, or for the zeal with which he improved the soil and surface of his land. " He considered it patriotism to convert one blade of grass 'into two,' and so well were his efforts approved that he was made President of the Derbyshire Agricultural Society for the year 1828."

Chapter XV.

THE MARPLE LIME-KILNS

(*By* A. HULME)

A FEW minutes' walk along the road from Posset Bridge in the direction of Strines brings us to what would appear at a first glance to be the ruins of a mediaeval fortress. These strange-looking buildings are the remains of Oldknow's lime-kilns. The high wall in the front of the kilns was strengthened and its line broken by circular buttresses, faced so as to imitate an ancient castle on account of its commanding position as viewed from the opposite slopes of the valley. These kilns, situated on the banks of the Peak Forest Canal, commenced working in the autumn of 1797, when the summit level of the canal was completed.

One of the main objects of the canal, mentioned in the Act of 1794, was the direct, cheap, and easy means of procuring limestone, both for agricultural and building purposes, and with the opening of the canal the time was propitious for Oldknow to commence this business.

The original cost of the kilns was £1,527 10*s*. 9*d*. A further sum of £3,033 13*s*. 0*d*. was spent on additions and improvements, which probably included the cost of making the arm of the canal to Posset Bridge. In all the total capital outlay was £4,561 3*s*. 9*d*.

The earliest consignment of limestone by the Peak Forest Canal of which there is any record was on July 31st, 1797. This fact, considered along with a memorandum dated the 11th November, 1797, to the effect that J. Wardle agreed to boat the limestone from Bugsworth to the wharf at Marple and discharge the same at 5*d*. per ton, seems to show that this business commenced during the autumn of that year.

The limestone was conveyed from the quarries at Dove Holes and Barmoor down the Peak Forest inclined plane tramway to the canal basin at Bugsworth and loaded into the boats. On arrival at Marple it was tipped into the kilns along with the coal drawn from the pit conveniently situated on the spot. The finished product was then taken out of the kilns from the tunnels below, into which the wagons ran, and conveyed down to the boat house, where there was accommodation for two boats under cover from the rain, or conducted into another building and loaded into carts secure from the weather. Mules were used for drawing the wagons backwards and forwards along the tramway. Before the opening of the locks the tramway ran down to the bottom level at the "Grand Aqueduct"—as it was termed in those days—a distance of about $1\frac{1}{4}$ miles.

Early in 1798 Oldknow agreed to let his lime burning to Joseph Heathcote, Senr., Joseph Heathcote, Junr., Wm. Fletcher, and Anthony Ashmore, and to pay them 2/– for every score of horse loads produced, but if at any time he reduced his price to 10*d.* per horse load he would reduce their rate of pay to 1/8 a score. There is no indication that Oldknow reduced the price of lime to his customers, but in 1799 he certainly reduced the rate of pay of the lime burners to 1/8 a score. By the summer of 1798 a new kiln had been finished, and in the afternoon of July 7th it was filled and lighted. At this time carters were conveying lime from the Marple kilns to the canal bank on Lancashire Hill, Heaton Norris, and receiving 5*d.* for every horse load so conveyed.

This minor industrial enterprize of Oldknow's appears to have suffered a gradual decline from 1805 onwards, which was probably due to the Napoleonic War. Oldknow's complaint in 1806 of "the high price of labour and materials"[1] is significant of the times and goes to confirm this inference.

[1] In his letter to the Bishop of Chester. *See* p. 234.

In 1799, 52,000 loads of lime were produced, the wages bill for the whole of the twelve months amounting to £378. This included the cost of unloading the boats, stacking the limestone on the wharf, and discharging coals and limestone into the kilns. According to the canal permits for this year, the actual cost of the limestone purchased from the Canal Company at the bottom of the tramway at Bugsworth was 11*d.* per ton. In addition to this there was the cost of boating and the waterway toll. This toll was fixed by the Act of 1794 at 1½*d.* per ton per mile (a fraction of a mile to count as one mile). The actual distance from the basin at Bugsworth to the Marple kilns being 6 miles, 39 chains, would give a tonnage rate of 10½*d.* Yet during this year for some unexplained reason we find that Oldknow was charged 1/– per ton canal toll. The following year the output of the Marple kilns had increased to 74,000 horse loads, and the wages bill to £530. This increase in wages was not only due to increased output, but also to the building of new kilns. In his factory Oldknow had different rates of pay for the same occupations, and this feature is also noticed in the pay of the lime burners. During 1800 J. Heathcote's gang were paid 1/8 a score loads for burning, and the other gang, N. Ashton & Co.—working during the same period—were paid 1/6 a score loads. In the ensuing year they are both paid at the same rate, viz., 1/8.

The records for the year 1801 are incomplete. Three months' output, however, can be traced :

March	3,160 horse loads.
August	6,600 „ „
December	1,080 „ „

In December, there being insufficient work at the kilns, the lime burners are put to work on the farms.

The big decline in the winter was probably due to the freezing of the canal, and the dearth of building operations in frosty weather.

At this time there is nothing to indicate that Oldknow's lime business was not a paying concern. That he intended to develop still further this industry is fairly evident, for during 1802 cottages were erected close to the kilns for the housing of those working thereon. The output for this month is the largest recorded to date, viz. 13,420 loads, and without doubt the new kilns were working.

Half of the lime produced at these kilns was taken by water as far as Bolton, Bury, Rochdale, Saddleworth, and other Lancashire and Yorkshire towns. Of the other half, about three-quarters was used for agricultural purposes, the remaining quarter being absorbed by local bleaching houses, calico printers, and building operations.

The cost of the limestone had risen by 1803 to 2/4½ per ton, at which figure it remained for the next five years. In 1808 a slight reduction was made to 2/2 per ton, but by that time Oldknow was working the kilns at a loss. A number of monthly statements of expenditure and receipts have survived, which although incomplete show clearly the gradual decline in the profits of the lime burning. The expenditure includes the cost of the limestone, boating, coals, wages, and the canal tolls. These items for four months in 1804 amounted to £1,856, and the receipts from the sales of the lime and lime ashes for the same period brought in £2,755, leaving a profit of £899 for the four months. The same period in the following year shows a decline in the profits of £207, and the account for ten months in 1806—the most complete record we have—shows a profit of only £549. These figures do not take into account the interest on the capital outlay or debts, nor the upkeep of the kilns, the road, tramway, and canal connections. In addition to these there were the rates and taxes, which included window taxes, police rate, bounties for militia men, poor leys, highways rate, etc. Taking these into consideration there is little doubt that Oldknow was already losing considerably on this undertaking.

A few words on the wages question will help us to understand the position better. It must be remembered that at this time a European war was raging, which, as in the recent war, brought about a gradual increase in the cost of living, with a corresponding rise in the cost of materials and labour.[1] In 1804 (February 4th to March 3rd) to produce 7,647 loads of lime cost £89 in wages. A similar output in 1805 (February 2nd to March 2nd), 7,744 loads, cost £119, and by the end of 1806 (November 8th to December 6th) 7,438 loads cost £154 in wages.

Thus it will be seen that the wages bill alone had practically doubled itself in just over two years.

Early in 1807, Oldknow, realizing that his enterprize was on the decline, endeavoured to improve the business by advertising, and during February we find him paying C. Wheeler, the Manchester publisher and printer, the cost of such advertisements. The lime-works cash-book at this period records the price of newspapers at 6*d*. each.

[1] Examples indicating increases in rates of pay per day of Oldknow's outdoor workers:

	1799	1800	1801	1802	1803	Occupation
William Ashton	—	—	3/–	3/–	3/6	Building.
Samuel Barton	2/2	2/4	2/8	2/8	3/-	Road making, etc.
Thomas Batley	—	2/10	—	3/–	3/6	Building.
Joseph Rose	2/6	2/10	2/10	3/–	3/6	Building, Quarrying, Farming, etc.
John Day	—	2/10	3/–	3/–	3/6	Building.
George Fletcher	—	1/8	2/–	2/4	2/8	Farming.

Although the last monthly statement of expenditure and receipts we possess is dated February 1st to 21st, 1807, we know that the process of attrition gathered momentum, and by 1808 the accumulated loss on this business totalled £2,027. Oldknow's lament that "the war has as it were swept away the means of effecting great designs"[1] can be readily appreciated.

The charge of 5*d.* per ton for boating, 1797–8, had risen by 1806 to 7*d.* but the effect of this increase was more than counteracted by a reduction in the canal toll on the limestone from 1½*d.* a mile to 1*d.*

The lime-works cash-book which covers a period from July 1806 to September 1808 contains many interesting entries. From this we learn that the office was broken into by thieves on March 21st, 1807, and again in April 1808, and over £20 in cash stolen.

In addition to the coal drawn from the colliery on the kilns, Oldknow worked another shaft situated on the canal bank in a small wood known as Picky Wood, 200 yards beyond Brick Bridge, to which was a short arm of the canal. For the use of the canal from this pit to the lime kilns—a distance of just over 500 yards—Oldknow paid the Peak Forest Canal Company 1*d.* per ton toll. Now the canal accounts show that Oldknow was also procuring coal from a pit known as "Mancum," for which he paid the Canal Company 3*d.* per ton during 1808. The tonnage rate of 2*d.* per ton per mile for coal had been reduced by 1806 to 1½*d.*, and taking these facts into consideration we are enabled to fix the position of "Mancum" pit at a distance of two miles from the Marple kilns in the Disley direction. A forgotten and disused coal shaft situated on the canal bank 374 yards beyond Stanley Hall Bridge is without doubt the pit referred to, the distance from Oldknow's higher lime kiln

[1] Draft of letter to the Bishop of Chester, June 21st, 1806.

branch being two miles. This coal was of an indifferent quality known as "Sleck," for which Oldknow paid 5/– per ton. In charging the kilns, the stone was broken into lumps weighing from ½lb. to 6lbs. each ; three tons of limestone with one ton of coal produced on an average 15 loads of lime.

As the lime burning was being run at a loss Oldknow decided to give up this business, and he leased his lime kilns, canal basins, tramway and private roads connected with the same, to the firm of Wright and Brown. Although the proposals for taking over this business are undated, they clearly belong to the year 1811, for at this time Oldknow ceased paying rates on the kilns. No further references to lime burning appear in the records after this year, and the clerk at the lime office, Richard Goodwin, was found work in the new warehouse situated on the canal bank above New Marple.

Chapter XVI.

ROADS AND BRIDGES : PEAK FOREST CANAL : CHURCH BUILDING

(*By* A. HULME)

I.

ONE of the first undertakings of any magnitude in developing Oldknow's Marple and Mellor estates was the construction of the handsome stone bridge over the river Goyt close to the site where he intended building his factory. The only documentary evidence of this appears in a letter from William Strutt, Junior, of Derby, dated March 1st, 1789, in which he enclosed the plans for the structure. Strutt suggests that before Oldknow finally adopts this plan he should consult "some learned mathematician," and goes on to relate how "a man has got a patent for making Iron Bridges cheaper than of stone and has prepared a plan for one over the Schulkill in America,[1]—but I suppose you will not wait. Indeed, if we are to wait for the perfection of every proposed improvement, this age is so improving, we should sit still and do almost nothing." This bridge, which has a span of fifty-four feet, must have been finished by 1790, to afford those engaged in the building of the mill a quick and effective means of transport across the river. The next step in the development of communications was the making of the road up the face of the steep cliff known as "Goit Cliff." This involved—if we are to believe Oldknow's own statement—the cutting through of between eight and ten thousand square yards of solid rock. This road, which is about 500 yards long, and the building of the bridge cost not less than £3,000.

[1] The bridge referred to by Strutt was the invention of Tom Payne, who took out a patent in England in 1788. *See* Smiles' *Life of Telford*, p. 172.

A number of printed toll-bar tickets dating from 1795 to 1799, endorsed "Marple New Bridge Bar," indicate that a toll-gate existed on the Marple side of the bridge, probably on the site now occupied by Marple Lodge, where John Clayton afterwards took up his residence. The revenue derived from this toll-gate amounted to about £50 per annum.

The main road from Marple to Stockport at that time was very little better than a pack-horse track, and Oldknow, anticipating that his enterprize would be severely handicapped, commenced to repair or build in 1790 two bridges, one over the stream at the bottom of Dan Bank, and the other over Fogg Brook, Offerton. In 1792, when the Mellor mill was approaching completion, he had the roads connecting them repaired and gravelled.

As soon as the mill was finished and at work he turned his attention seriously to developing communications and transport, and became the chief promoter and chairman of the Peak Forest Canal. On December 5th, 1793, the Canal Committee was summoned to the Ram's Head Hotel, Disley, to meet the agent of the Duke of Devonshire, who must have offered no opposition of any consequence as the Act authorizing the construction of the Peak Forest Canal was passed on March 28th, 1794. This scheme was promoted and the Act passed during a "canal boom" of a similar character to the "railway mania" half a century later in 1845. In four years, up to the end of 1794, over eighty Canal and Navigation Acts were passed by Parliament authorizing the expenditure of more than £5,300,000. This mania for canal building was followed by a panic and many projects failed, but the Peak Forest Canal, after passing through many difficulties, opened on May 1st, 1800, with the exception of the Marple Locks.

The 1794 Act states that in carrying out the project

the proprietors were not to take, use, or damage any buildings or gardens in existence before January 1st, 1794, with a few exceptions. Two of these are: "A garden situate in the township of Marple belonging to John Taylor and another occupied by Thomas Sykes belonging to John Hollingworth."

The Canal Company had the option of adopting either or both of two routes near Marple Dale, and the route selected is marked by the Marple Aqueduct, which was considered one of the greatest engineering feats of the day. George Borrow, in his *Wild Wales*, refers to it as follows:

> Few things are so beautiful in their origin as this Canal, which be it known, with its locks and its aqueducts, the grandest of which last is the stupendous erection near Stockport which by the bye filled my mind when a boy with wonder, constitutes the grand work of England, and yields to nothing in the world of the kind, with the exception of the great canal of China.

This aqueduct was generally described in the Canal Carriers' accounts as the "Grand Aqueduct."

The Canal Company were not allowed to take any water from rivers near the canal except in times of flood. Two large reservoirs were constructed to act as feeders, viz. Todd Brook near Whaley Bridge and Combs reservoir near Chapel-en-le-Frith. They had power to take water from mines to the mutual advantage of each, but mines could not be worked to the prejudice of the Canal Company.

A connection through the Ashton Canal at Dukinfield Junction with the Rochdale Canal at Manchester enabled Oldknow to have direct water communication with many Lancashire towns, and through the Aire and Calder Navigation linked him up with the Yorkshire manufacturing districts and the river Humber. In the westerly direction he had a cheap and easy means of transport to and from Liverpool over the Bridgewater, the Rochdale and the Ashton Canals, the freight charges on 50 bags of cotton from Liverpool to Marple costing him only £3 10*s.* 6*d.* By the Ashton, the Rochdale, and the Bridgewater Canals there was likewise a connection with Brindley's Trent and Mersey Canal. Although a portion of the Peak Forest Canal opened during the year 1797, the locks connecting the top level with the bottom level at the aqueduct were not completed until 1804. In the meantime an inclined plane tramway supplied the place of the locks. Several canal permits which have been found show that large consignments of ten-inch paving stones were being conveyed from Bugsworth to Manchester during the year 1800. All this meant, of course, that a considerable amount of time and money would be spent in transferring many tons of stone from the boats into the wagons and again reloading into boats at the bottom level. On the completion of these locks, according to local tradition, Oldknow was very anxious that one of his lime boats should be the first to make the journey, and to encourage his workmen, who were engaged in building the bridge at the junction of the arm of the canal from the lime kilns, to make greater efforts he regaled them with ale posset. Tradition says he succeeded in accomplishing his desire, and that his boat dubbed the "Perseverance" won the day. This bridge is still known locally as Posset Bridge.[1]

Between the ninth and tenth lock above the aqueduct Oldknow erected a large warehouse, with which his factory was connected by his road up the Goyt Cliff and through New Marple.[2]

On the top level of the canal, close by the lime kilns, Oldknow had a boat building yard and dry dock, and here

[1] Wainwright's *Memories of Marple*, p. 10.

[2] New Marple was the name given to that part of the village comprising Stone-row and Brick-row (now demolished).

he built and repaired the boats that plied this waterway. The following are the names given to a few of them : "Marple," "Peak Forest," "Buxton Farmer," "John Bull," "Mary," "Mellor," "Quaker," "Rose," "Success," "The Duke," "True Blue," "Willersley Castle," "Windsor Castle," "Honesty." The carrying capacity of these boats was 25 tons each. Owing to the increase in the price of materials caused by the war the cost of building boats rose to £140 each.

Priestley[1] and Phillips,[2] the early authorities on the history of inland navigation, state that the canal was opened on May 1st, 1800, and neither of these two writers appears to have been aware that limestone was being conveyed from Bugsworth to Marple by canal as early as July 1797. The present writer has a canal permit in his possession dated July 31st, 1797, showing that 23 tons of limestone passed between these points. In all probability the tramway from the limestone and gritstone quarries and the summit level of the canal from Bugsworth to Marple would be finished first, to enable the engineer responsible for the undertaking to obtain stone for the building of bridges, locks, wharves, and warehouses, and also limestone for the macadamizing of the towing paths. In making the tramway the Canal Company cut into a bed of hard gritstone about half a mile from the terminus of the Canal at Bugsworth. It was soon found that this was a good wearing stone, and a large demand sprang up for paving setts which, combined with hard wear, had the advantage of never becoming slippery after wearing smooth.[3] We find the Court-leet

[1] *Historical Account of Inland Navigation and Railroads.*

[2] *General History of Inland Navigation.*

[3] Article published in the *Great Central Railway Journal*, December 1905, on "The Peak Forest Canal Tramway," by George Taylor.

of Manchester, in 1798, asking Oldknow to supply their "Paver" with seven or eight loads more of his stones and thanking him for the trouble he has taken.

A document discovered amongst Oldknow's papers dated December 17th, 1794, shows that he originally held 52 shares in the Peak Forest Canal Company. At this date a call of £8 per cent. was made on the subscribers, which along with their original deposits and first call made a total of £15 per cent. on each £100 share up to date. Of the £90,000 directed to be raised by the 1794 Act only £80,600 was subscribed, and not the whole of that sum had been paid. They had also power to borrow a further sum of £60,000 on mortgage of the rates, but were only able to obtain £36,540 in this way. By 1800 the funds were completely exhausted, and a further application was made to Parliament for power to raise, either by new shares or by promissory notes, any sums necessary to complete the undertaking, provided the whole should not exceed £150,000 including the previous subscriptions. Oldknow now increased his holdings from 52 hundred-pound shares to 261. This number included 47 shares purchased from his neighbour Miss Shaw, and also a number bought from a Mr. Thomas Winterbotham.

Notwithstanding this later Act of 1800, financial difficulties still dogged the proprietors, and at a general meeting held at the Red Lion Inn in Heaton Norris on November 19th, 1801, with Oldknow in the chair, the committee secured the consent of the shareholders to a proposal to make a call of an additional £10 per share, which was to be repaid with interest before any dividends should be paid.

The construction of the Marple Locks, 1803–4, and other expenses connected with the enterprize had absorbed nearly all the money raised by the last Act of Parliament, and only £573 5*s.* 0*d.* remained unexpended. In 1805 another Act empowered them to raise a further sum of £60,000 by creating new shares, at such prices as the proprietors thought fit.

At the time of making the Peak Forest Canal it appeared that a great advantage would be reaped by connecting the top level with the Trent and Mersey Canal, and that, by this short cut across Cheshire, Manchester and the rapidly growing industrial area around would have a more direct means of communication with the Potteries and the manufacturing districts then springing into activity in the vicinity of Wolverhampton and Birmingham. In 1796 a survey was made by Mr. John Nuttall of a line for a canal from the summit level of the Caldon branch of the Trent and Mersey Canal at Endon to the top level of the Peak Forest Canal at Marple, then in course of construction. Mr. Benjamin Outram revised Mr. Nuttall's survey and made a rough estimate and report, which was printed March 10th, 1796. A Bill for the construction of this waterway actually passed the House of Commons but was thrown out of the House of Lords by the influence of the Duke of Bridgewater and the contending interests of various landowners and canal companies. It was the intention of the promoters of this scheme to make a branch canal from Marple to Stockport, where it would have approached the Heaton Norris branch of the Manchester, Ashton, and Oldham Canal. In 1825, however, another effort was made to obtain powers to connect these two waterways, when the promoters obtained the services of the renowned Thomas Telford as the engineer, and on April 11th of the following year royal assent was given to the company's Bill.

It is not surprizing to find amongst the names of the proprietors that of Samuel Oldknow, as the benefits that would be conferred by this new waterway on the small industrial community he had founded in Marple can be readily appreciated when it is considered that the distance to the Midlands and London would be lessened by 40 miles. The Macclesfield Canal was not completed at the Marple end till a year after Oldknow's death.

II.

During the period of the Peak Forest Canal construction Oldknow was by no means unmindful of the importance of road communications. Towards the end of 1796 meetings had been held between himself and two other principal landowners in the neighbourhood to consider the question of making a turnpike road from Stockport through Marple and New Mills to Hayfield, and in January 1797 a plan was agreed upon with Wilbraham Egerton, who owned an estate on the banks of the Goyt near Strines, and Thomas Beard, a New Mills landowner. But Oldknow's propensity for changing his mind once more asserted itself. A few days after the meeting in question, on January 21st, he wrote to Egerton informing him that he intended to depart from the original plan and to make the road to Marple Bridge only. This, however, did not suit Egerton, as the projected road would not pass through his estate, and on February 8th he wrote promising Oldknow every opposition in his power when his scheme came before Parliament. Egerton's remonstrance had the desired effect, and the Act procured was for a turnpike road to New Mills and Hayfield including a branch to Marple Bridge.

Early in 1798 Mr. Thomas Brown, who had superintended the construction of the Peak Forest Canal, made a preliminary survey of the existing road from Stockport to Marple Bridge. Oldknow's original proposals were that his road down the Goyt Cliff and over the stone bridge close to his residence should be included in the intended turnpike to New Mills, provided that £1,000 or the interest on that sum at 5 per cent. be secured to him in the intended Act. In a letter to Mr. Thomas Beard he points out that he has already spent £3,000 in building his bridge and making "an almost unparalleled costly road," and that "when it is considered that the money is already expended—and the present Tolls amounting to £50 a year—I should, by a further accommodation to the public and cuting through my home-stead, put myself in a worse situation than I now stand were the interest left uncertain."

He then goes on to suggest that this proposed interest should be paid him half-yearly, before any other dividends were paid, out of the tolls arising from one toll-gate to be set in Offerton and another in Whittle, and on these terms he would undertake to solicit subscriptions of all the interested persons from Marple to Stockport, if Mr. Beard would undertake the solicitation of all interested parties from Marple to New Mills, Thornset, and Hayfield.

During September 1798, Oldknow, along with Thomas Brown's surveyor, was called upon to attend a meeting of the Road Commissioners in order to "shew them the advantage of coming down what is called Mr. Oldknow's road in preference to Turncroft Lane." Oldknow's proposals that his private road should be included in the intended turnpike, for some unexplained reason, did not mature, and the new road at this point was projected round the shoulder of the hill instead of descending into the valley. It then follows a course almost parallel with the canal in the direction of New Mills.

Much progress must have been made with the new turnpike in the Marple area by 1802, for during January and several months following a number of Oldknow's men are recorded as working on the "New Turnpike Road on Mr. Oldknow's Account," and in April of the same year Ralph Hammond is "Fencing Stockport Road," his rate of pay being 2/8 a day.

In July the men are found at work on the old highway leading from Norbury Smithy to Disley. This ancient highway, which previous to the making of the road referred to was the main thoroughfare through the village, was much too narrow in parts and had been sadly neglected. Oldknow now proceeded to widen those portions necessary, and also diverted the road round the east side of the ruinous old church to the west side, cutting off an awkward curve and shortening the distance.

A letter, undated but obviously belonging to this period, written by a Marple landowner suggests that Oldknow's bridge over the Goyt was closed on Sundays.

"for Mr Samuel Oldknows Cotten Manufacturer

"MARPLE, *November*.

"SIR,

"I hope to give no offence I want to Buy a favour of you, that is to go on foot over your Bridge on the Sabbarth Day to Mellor Church. I'le give you Some Money Every yeair. If it will be Granted It is the first favour I ever asked you and the first you wanted of Me I Granted —what is that you'l say—why to take a corner of My field at the End of your New Road. If this favour will be Granted I will Wait on you at Mr Bennets' or R. Coliars To Treat with you for it.—

"from your friend

"OTTIWELL HEGINBOTHOM."

III.

"Gentlemen, I have made you excellent roads upon earth, and [pointing to the Church] I have made you an excellent road to Heaven." The well-attested tradition of this speech of Oldknow's at a vestry meeting to mark a victory in a dispute with the surveyor of highways is a sufficient justification for adding to this chapter a brief account of his church-building at Marple. Like most of the leading industrialists of his time, Oldknow was brought up a Dissenter. Rivington Chapel, where his father was buried and where he himself was a worshipper in Anderton days, belonged to that least orthodox of religious communities whose congregations, known as English Presbyterians in the middle of the 18th century, frequently adopted Unitarian doctrines in the days of Priestley. Whether he joined the congregation at Stockport which was moving in that spiritual direction is unrecorded, and it is also uncertain how far his initiation as a Freemason at the Lodge of Benevolence in 1787 may be taken as evidence of his persistence in the paths of liberal theology. A more probable explanation of this step is to be found in the strength of his social instincts and his passion for building, and these motives no doubt played a part in the rebuilding of the chapel of the Established Church at Marple. The old chapel was a black-and-white barn-like structure, built two centuries before and much too small for the new community, whose spiritual needs were being provided for by the Methodists. There were 130 children in their Sunday School in 1797, in 1798 they acquired a house at Norbury Smithy belonging to Joseph Rowbotham and registered it as a place of worship, and by 1804 they had built a chapel on land bought from Ottiwell Heginbotham for £25.

These activities no doubt stirred the supporters of the Established Church to emulation. In 1803 the incumbent, the Rev. W. Bowness, B.A., inspired by Oldknow and supported by the churchwarden, Mr. John Mason, started a movement for the rebuilding of the chapel and a sitting was offered to every subscriber of £5. The appeal evidently met with a poor response, and on June 21st, 1806, Oldknow addressed the following to the Bishop of Chester :

" My Lord,

" I make haste to inform your Lordship what is *done* and what is *not* respecting Marple Chapel—We have obtained Land to make a most commodious Chapel yard—and the same is walled with an Excellent Stone wall on 3 Sides thereof and the Public high way is turned from the East side to the west side of the yard—but the Old Chapel remains in ye same ruinous state it has done for 3 or 4 years—not repaired—not Rebuilt—When I had last the honour of speaking with your Lordship at Mr Marklands on this subject we then wish'd for consent to remove the Chapel on to new Ground now a part of the enlarged yard—to this Mr Marriot would not give his consent—We therefore resolved to rebuild & enlarge the Chapel on the present site—and I have twice consulted Mr Web on the plan which is concluded upon—& we wait for nothing (but a Dread of the Money being too Little) and an opportunity of making it more by Soliciting Lord & Lady Bulkely who possess the valuable presentation of the Mother Church (Stockport) & so soon as I hear of their arrival at Pointon I will wait upon his Lordship with our Subscription paper—which amounts

to about	600	0	0
and add Mr Oldknows speculation for the surplus seats	400	0	0
& left of a ley collected to repair but which was thought inexpedient	100	0	0
	1100	0	0

"A sum sadly too little to build a Decent Commodious Church particularly when we take into the account—the high price of Labour & materials.

"Besides my Lord the war has as it were swept away the Means of Effecting great designs. Give us peace & the Church shall not be 3 months unbuilt."

The old chapel described in the preceding letter was pulled down in 1808, and presumably the building of the new church commenced the same year. The following year (1809) we find several of Oldknow's workmen engaged in the erection of the edifice, and timber required for the same is being brought from Hyde Bank, Windy-bottom Wood and Cravens Wood.

The total seating accommodation in the new church was 552. Of these 325 were held by Oldknow, and if the appeal of September 17th, 1803—in which £5 is the subscription required to secure one sitting—is a satisfactory guide Oldknow must have subscribed at least £1,625 towards the new structure. The total number of seats in the old chapel was 172, and these were held by the inhabitants by presumptive right, but only 76 of them were claimed in the new building. At the request of the Bishop of Chester that certain seats should be allotted for the poor, Oldknow observed that the poorer class of inhabitants were his own workpeople, for whom the surplus seats in the gallery were intended.

Oldknow died at Mellor Lodge on September 18th, 1828, and he was buried on September 24th at Marple Church. "Few men," says the *Gentleman's Magazine* of November 1828, "who have of late quitted this transitory scene have led a life of greater industry and more active benevolence, or died more universally lamented than this individual . . . How he was loved and honoured is perhaps best told by the spontaneous feeling of all classes of society on that occasion. From an early hour the people began to assemble, and lined the way from his house to the Church, closing as the procession moved along. On its arrival at the gateway a line was formed by the children from the Military Asylum, each dressed in a scarlet spencer, and a black band round the arm . . . The funeral service was read by the Rev. Mr. Litler. The Reverend Gentleman himself was much affected and hundreds gave free vent to their feelings of real sorrow . . . Probably the number assembled was not less than 3,000. As it was the general wish to see where the body was deposited, several hours elapsed before the vault was clear."

FIRST METHODIST CHAPEL, MARPLE

Chapter XVII.

SAMUEL OLDKNOW: THE MAN AND HIS WORK

Landlord Edmund was seen and felt by all men to have done verily a man's part in this life-pilgrimage of his ; and benedictions, and outflowing love and admiration from the universal heart were his meed. "Well done ! well done !" cried the hearts of all men. . . . .

All this that thou now seest and namest Bury Town, is properly the Funeral Monument of Saint or Landlord Edmund. . . . .

Plugson, who has indomitably spun cotton merely to gain thousands of pounds, I have to call as yet a Bucanier and Chactaw ; and till there come something better, still more indomitable, from him.

—CARLYLE : *Past and Present.*

THE career of Samuel Oldknow as a man of business may be readily summed up in the statement that he gained a fortune as a manufacturer and lost it as a spinner. But even from the purely economic point of view a record of failure may have as much scientific interest as one of success ; and it is as easy to over-estimate Oldknow's technical achievement in the period of his business success as it is difficult to appreciate with adequacy the extent of his social achievement during the much longer period whilst he was scattering, as seed corn, the harvest he had reaped.

Our records not only confirm the tradition that Oldknow was the first successful maker of British muslins, but seem to show that he owed his success more to creative aesthetic gifts than to mere business shrewdness. It is not impossible that, if all the facts were known, we should have to attribute greater merit in the origination of the manufacture to Thomas Ainsworth of Bolton, to Joseph Shaw of Anderton, or to some " mute inglorious " weaver of Paisley who struggled in vain against the more adverse conditions of an earlier time. To such a pioneer Oldknow would stand in the same relation that Richard Arkwright bears to Lewis Paul or George Stephenson to Richard Trevithick, except that his leadership in the new field, though brilliant, was far briefer. From first to last his career as a muslin manufacturer covered less than a dozen years, and before half that time was spent he had begun to transfer his energies to the task that was to occupy the remaining half of his life.

If the reader seeks Oldknow's monument he should go to Marple and look about him, setting out from the Weir Floodgates[1] and passing between the present course of the river and the great reservoir which represents its old course, to the solid farm buildings and Apprentice House at Bottoms Hall, and then by the side of the dried-up Mill Pond to the ruins, now more impressive below ground than above, of Mellor Mill, and past Mellor Lodge—built partly on arches above the river—over Oldknow's bridge and up his rock-hewn road, to the group of houses built for the first industrial colonists, to Posset Bridge, the picturesque lime-kilns, and the Church ; and finally passing down the sixteen canal locks to the great aqueduct that spans the Goyt Valley. And yet, deeply impressed as he will undoubtedly be by these visible survivals of Oldknow's life-work, he will have missed its true inward purport unless by some chance wayside inquiry he opens the springs of the still living tradition not merely of the Founder or of the Master, but of the Man.

Of Samuel Oldknow the man the reader who has followed his fortunes so far in these pages cannot fail to have a definite impression which a few final reflections will not much alter or improve. Yet something must be said of him for which this is the appropriate place. We know nothing of Oldknow's education, but in the intellectual and spiritual environment of his early manhood he had some share of the best that England then afforded. At Rivington Chapel he " sat under " a minister who had

[1] See map at end.

been trained at Warrington Academy, which, in the age that heralded the Revolution, was the chief seedplot in England of scientific discovery and of political and philosophic liberalism. Of Priestley, a former tutor at Warrington, who had now become the guide, philosopher, and friend of Boulton, Watt, and Wedgewood at Birmingham, and was soon to be the "patriot, saint, and sage, loved and honoured almost profanely" by the youthful Lamb and Coleridge, he must have heard much; and some echoes may have reached him of the gatherings of the Birmingham circle in the Lunar Society, which was opening a new epoch in the relations of science to industry. Of more direct and regular contact with this new learning the only evidence we possess lies in the correspondence with William Strutt already cited, but Oldknow's eager interest in all the practical applications of science has been shown in the later chapters of this book, and the last letter of his that has survived, written when he was nearly seventy, is to thank a Richard Ainsworth of Bolton for calling his attention, as a member of the Cromford and High Peak Railway Committee, to the superiority of malleable over cast-iron rails.

Of another—perhaps the most creative—of the spiritual influences of his time he was unusually receptive. The pianoforte, which has done so much for the diffusion of musical culture, was still in process of invention. Erard made his first English piano in 1780, but many fundamental improvements were still to be added by Stoddart, Broadwood, and others. It is likely enough, therefore, that the instrument installed at Heaton in 1786 was one of the first to be heard in the North of England.

"I have sent you," writes Salte on December 15th, 1786, "the Piano Forte, packed up very Carefully by Pickford's wagon—I hope you will find sweet music in it that will wrap your Soul as Milton says in Elizium.

Get the Sweet Nymphs about you, they will waken it and charm your Ears and mind with sounds of sweetest Melody." It was in this year, the reader may be reminded, that the world was enriched by the immortal music of *Figaro*.

The rest of Salte's exhortation leads us to the consideration of another aspect of Oldknow's personality. "Live for some fair Lady. Direct your attention to that State where true Felicity is only to be found. Dull, wretched Batchelors cannot be said to Exist. Solitary enjoyment deserves not the name of Pleasure. However you are half persuaded to become a Xtian. The Altar I hope will soon receive your Vows and the Vows of the Woman you already think an Angel."

It was not perhaps to be expected that our records would solve the mystery of Oldknow's love affairs, but they certainly add a not unpleasing degree of complication to them. The local tradition of a devoted attachment to his neighbour Miss Shaw (Wainwright: *Memories of Marple*), whose ill-health frustrated their union, reveals a romantic conception of Oldknow's character which appears at first sight totally inconsistent with the story of his engagement to Miss Drinkwater as related by Robert Owen. But it must be remembered that whilst Owen's word sufficiently vouches for the mere fact of the engagement, his attribution of mercenary motives is confessedly borrowed from common talk, and is obviously coloured by Owen's natural resentment. A withdrawal on the father's side for financial reasons would reflect no dishonour on Oldknow, and it is even possible that the discovery that the lady's affections were fixed elsewhere may have played a part in the matter. The affair of Miss Shaw, if based on fact, belongs to a later period when Oldknow was resident at Mellor.

The letters of Salte, Arkwright, and Strutt seem to

afford materials for an earlier chapter of romance. Salte's recurrent exhortations to marriage might be taken as evidence of nothing but his own exuberant nature, but the passage last cited certainly suggests confidences made to him during his recent visit to Heaton, and it is difficult to avoid connecting with this the interpretation placed by Arkwright in June 1787 on Oldknow's two recent visits to Derby. "There is certainly something going on between Miss S. and you, and of a very serious nature too. . . . So I think." And in September 1787 he returns to the topic : "They say Mr. Macmurdo and Miss Strutt are on good terms."

All this might be treated as "vacant chaff well meant for grain" if it were not for two of William Strutt's letters to Oldknow. The first of these, dated September 1st, 1788, deals with the elder Strutt's intention to take up bleaching, and thanks Oldknow for his offer to show father and son over some bleachworks, and adds : "There is a great commotion here, in people's minds at least, occasioned by the expectation of the great musical festival such as has not yet been at Derby before—every house is to be full. I wish you could come—the town will certainly be *very* gay but that is no inducement to you—your hobby horse is of another kind. Farewell & believe me Your affectt. friend." The second letter is dated August 16th, 1791, and, after dealing with various other matters, concludes : "I suppose you have heard that my sister is at last married to Mr. Fox. It was time to make an end of it & the attachment had lasted too long and was too strong to be broken. They are now at Brighton & we are left a forlorn herd of Bachelors. What will become of us the Lord knows, Adieu." Whatever construction may be put by "Ladies who have intelligence in love" on these facts, it will probably be conceded that Oldknow was more to be pitied than blamed for his celibate existence.

The open countenance, the large and expressive eyes, the full and mobile mouth of Oldknow's portrait, reveal unmistakably a sensuous and impressionable temperament, a genial and sympathetic disposition. That he had the defects of these qualities—that he was far from a good man of business, that he often failed to realise his own sanguine expectations and to keep his promises to others—is shown by the frequent and lively remonstrances of his friends, but that he retained their friendship and respect through his worst misfortunes is sufficient evidence of their unshaken confidence in his character.

Perhaps the most convincing testimony, however, to his open-hearted simplicity and disinterested benevolence—which are the vital and creative qualities in the founder of a community—is to be found in the instinctive confidence in them implied in the letters of all who had personal contact with him—his uncle the tea dealer, his managers and workmen, his friends in the business world. To one of the latter, a Lancashire man living in London, Oldknow had written at the most desperate moment of his own fortunes in September 1797 to give some account of a young ne'er-do-weel whom the friend had committed to his charge. "If," replied the anxious father, "you can but find the means as you say to rivet the attention of my lad to any object of prudence, not very material to me what, I should be extremely happy . . . if you would but have the goodness to hear what he may have to say in his own vindication and also what you may hear to his prejudice, I am well convinced that you are capable of judging of and making allowance for the frailties of youth."

In spite of all the noble eloquence and sound prophecy of Carlyle, those who know something of both periods may be permitted to doubt whether the twelfth century was any better than the nineteenth. Abbot Sampson, the poor boy who rose to be a mill monopolist and a great magnate, and who "by impetuosity of driving or adroitness in leading generally remained master at last," had much in common with Plugson of Undershot. Doubtless, too, there have been many Landlord Edmunds of a later age whose merits lie undisclosed for lack of a devoted bard. If Carlyle had chanced on the records of Mellor as well as on Jocelyn's Chronicle he might have struck a juster balance between Past and Present.

APPENDIX A

LIST OF GOODS MANUFACTURED BY S. OLDKNOW, COMPILED FROM HIS COSTING BOOK

Muslins : Plain, Spotted, Striped, and Checks.
Romals : Light Ground, Dark Ground, Pink and Blue.
Cravats : Jaconets, Tan Jaconets.
Ballasores : Plain, Checks.
Handkerchiefs : Ballasore, Spotted, Check, and Bordered.
Aprons : Plain and Bordered, Stripe and Bordered, Vandyke.
Calico : Plain, Blue Check, Shirting, Sheeting.
Doosooties : Doyleys, Huckaback.

APPENDIX B

SPECIMEN COSTINGS, FROM COSTING BOOK

9/8 Callicoe, 54 Reed—	£	s.	d.
73 hks. Twist, No. 26	0	13	1½
3 lbs. 6 ozs. Weft, No. 28, at 3/7	0	12	1
Weaving	0	7	0
Winding 6*d*., Warping 2½*d*., Sizing 3½*d*.	0	1	0
Bleaching, Finishing, etc.	0	2	0
	£1	15	2½

7/4 Striped & Border'd Aprons, 62 Reed—	£	s.	d.
84 hks., No. 50	1	4	1
74 hks., No. 36	0	15	1
Sizing 8½*d*., Windg. 2/5, Warpg. 7*d*.	0	3	8½
2 lbs. 0 ozs. Weft, No. 69, at 13/10	1	7	8
10 ozs. Weft, No. 42, at 6/–	0	3	9
Weaving	1	9	0
Bleaching 4/8, Finishing 2/6	0	7	2
	£5	10	5½

APPENDIX C

PERCENTAGE OF GOODS MANUFACTURED, 1784–1794

	NON-MUSLINS			MUSLINS							TOTALS	
	Cal.	Shirtg.	Fust.	Plain	Hfs.	Stripe	Chks.	Aprons	Romals	Spot	Non-Muslin	Muslin
Mar. 1784. Ledger	48	..	..	5	..	5	40	..	..	2	48	52
Nov. 1785. Ledger	32	8	..	15	..	15	22	8	..	..	40	60
Nov. 1786 } Taking-in book	20	5	5	20	2	2	30	14	2	..	30	70
Dec. 1786 } Taking-in book	22	1	6	20	5	2	25	14	2	3	29	71
Mar. 1787 } Taking-in book	21	..	5	19	16	..	12	16	6	5	26	74
Dec. 1788. Warping book	21	..	..	30	30	..	9	..	4	6	21	79
July 1789. Pay Tickets	12	1	1	20	30	8	13	..	6	9	14	86
Nov. 1790. Ledger	15	..	..	13	15	42	..	..	6	9	15	85
Dec. 1790. Warping book	9	1	..	30	18	31	4	..	7	?	10	90
Dec. 1791. Warping book	38	1	..	9	20	14	16	..	2	?	39	61
Dec. 1793. Pay Tickets	45	1	1	18	13	8	10	..	1	3	47	53
June 1794. Pay Tickets	38	7	3	30	14	2	6	..	..	..	48	52

APPENDIX D

LIST OF PATTERNS WORKED UP IN THE FINISHING ROOM, 1788

(From Record of Finishing Frames)

Cups.
Cup and Ball.
Aprons
„ Spot.
„ Vandyke
Triangle, Double.
Pepper Box.
Club.
Chain, Single.
„ Double.
„ Hard.
„ Broad.
„ Painted.
Bromeleys.
Great Cross +.
Little „ +.
W. Pattern.
Red Stripe.
Truelove Knot.
Diamond.
Barley Corn.
Red Figure.
Narrow Pointed.
Spade
„ Hard.
„ Double.
„ Red, Soft.
Hailstones.
Needlework.
Open Work only.
Spot, Soft.
„ Small, Round.
Hollow Dice.
Dice.
Cockleshells
„ Opened.
Handkerchiefs.
Sprigs
„ and Firtrees.
2 Leaves.

APPENDIX E

SALES BOOK

	£	s.	d.
1786—June	5527	15	0
July	5397	3	0
Aug.	5993	13	0
Oct.	4469	17	0
Nov.	6007	15	0
Dec.	2453	9	0
1787—Jany.	4453	15	0
Feby.	3648	9	0
March	4804	16	0
April	5750	18	0
May	6323	0	0
June	4648	1	0
July	4988	16	0
Sept.	5584	11	0
1788—May	4399	0	0
June	4441	0	0
July	3132	0	0
Aug.	3580	0	0
Sept.	3639	0	0
Oct.	2504	0	0
1789—April	10391	0	0
May	10585	0	0
June	7979	0	0
July	9290	0	0
Oct.	8438	0	0
Nov.	4808	0	0
Dec.	3421	0	0
1790—Jany.	6010	0	0

APPENDIX F

OLDKNOW'S CUSTOMERS, c. 1789

S. & W. Salte, London.
Parker, Topham, & Sowden, London.
Greaves & Hodgson, London.
Joseph Jones, London.
Howarth & Smith, Manchester.
John Phillips, Manchester.
Osbalderstone, Manchester.
Smith & Blackwell, Manchester.
Sykes & Townsend, Manchester.
Holland & Busyard, Manchester.
Robt. Markland, Manchester.
Kersley, Manchester.
Taylor & Withington, Manchester.
John Robinson, Manchester.
C. F. Brandt (Agent), Manchester.
Reynolds & Warner, Birmingham.
Hayley & Fox, Birmingham.
Wm. Lycett, Birmingham.
Guest & Reynolds, Birmingham.
Wm. Hope, Liverpool.
Tarleton & Backhouse, Liverpool.
Sidebottom & Horrocks, Liverpool.
Taylor & Almond, Nottingham.
Fowler & Collingham, Nottingham.
Siddons, Mansfield.
Richard Arkwright, Cromford.
McGuffog, Stamford.
John Jackson, Leicester.
Brewin, Leicester.
Henry Hindley, Stockport.
Coppack, Stockport.
Bury & Dodge, Stockport.
R. Bunney, Coventry.
Thos. Goodwin, Coventry.
Parsons, Studley & Hurle, Bristol.
Gould, Bakewell.
Thacker, Wolverhampton.
R. Battelle, Derby.
Cooper & Freemantle, Sarum.
Robert Blackwail, Wirksworth.
B. Gibson & Son, Newcastle.
Owen Williams, Holywell.

Private Customers

Mrs. Smithells, Stockport.
Mrs. Taylor, Manchester.
Mrs. Langford, Manchester.
Mr. Foye, Bristol.
Mr. John Harwood, Bristol.
Mr. Chipchase, Durham.
Wm. Forman, Doncaster
Lord Penrhyn, Winnington.
Lady Warren, Poynton.

Foreign

Von der Becke, Leipsic.
Chaudoirs, Paris.
Tiberghien, Brussels.

APPENDIX G

PRICES OF WEAVING

Plain Ballasores

	34	38	46	50	56	60	64	68	72
35	3/7	3/10	4/6	5/1	..	..	..	..	..
36–39	..	..	4/10	5/5	5/11	..	..	..	..
40–43	..	..	..	..	6/4	6/9	..	..	..
44–46	..	..	..	..	6/10	7/3	7/10	..	..
47–49	..	..	..	..	7/3	7/9	8/5	..	..
50–52	..	..	..	..	7/8	8/3	9/–	..	..
53–55	..	..	..	..	8/3	8/10	9/6	9/9	..
56–58	..	..	..	..	8/10	9/5	10/–	10/5	11/–
59–61	..	..	..	..	..	..	10/6	11/–	11/7
62–64	..	..	..	..	..	..	11/–	11/6	12/2
65–67	..	..	..	..	..	..	11/6	12/–	12/9
68–70	..	..	..	..	..	..	12/–	12/7	13/4
71–73	..	..	..	..	..	..	..	13/2	13/10
74–76	..	..	..	..	11/3	12/2	12/11	13/8	14/5
77–79	..	..	..	..	13/1	..	..	14/3	15/–
80–82	..	..	..	..	..	..	..	14/10	15/8
83–85	..	..	..	..	..	..	..	15/4	16/3
86–88	..	..	..	..	..	..	..	..	16/9
89–91	..	..	..	..	..	..	..	..	17/4

APPENDIX H

RADCLIFFE'S "NEW SYSTEM" IN OLDKNOW'S STOCKPORT MILL

The careers of Samuel Oldknow and of William Radcliffe as captains of industry are so curiously interwoven that whilst it is impossible in this book to settle the many vexed questions that arise out of Radcliffe's rambling and controversial but profoundly interesting narrative, it is equally impossible not to give some account of the circumstances under which he took over Oldknow's group of buildings at Stockport and carried out the purposes for which they had been designed. Although Radcliffe was five years younger than Oldknow, and succeeded to his position as a large manufacturer ten years later, he had been his predecessor in Mellor as a spinner, a landowner, a roadmaker, and a prospective Lord-Lieutenant. Down to 1794 his chief business had been the production and sale of warps for Oldknow's weavers, whilst he gave out some work to muslin weavers on his own account. His conversion in that year to a predominant interest in weaving is attributed by himself to the revulsion aroused in his mind by the diabolical suggestion of a foreign buyer that he should spin yarn for exportation.[1] The historian may perhaps be pardoned if he connects the change with Oldknow's abandonment about this time of his manufacturing business, and with the influence of the " Golden Age " which, by Radcliffe's own account, was then at its height after the crisis of 1792–3.

Whether the motives of his decision were patriotic or economic, or a judicious mixture of both, the results down to 1801 certainly seemed to justify it. So great an export trade had grown up—of a kind thoroughly approved by his mercantilist conscience—" in plain and fancy goods for the Leipsic and other Continental fairs as to employ upwards of 1,000 weavers, widely spread over the borders of three counties, whose wages sometimes amounted to £700 for a week's work." [2] If Radcliffe's memory did not deceive him in making these estimates, his manufacture was at that time more extensive than Oldknow's had ever been, though his wares were not of so fine a quality and his profits probably not so great. He had taken as partner Mr. Thomas Ross, of Montrose, who brought in ultimately a capital of £6,000, and was better qualified than himself to conduct the sales and correspondence, to travel on the Continent, and deal tactfully with the hated but necessary foreigner.[3] The firm moved its headquarters to the Muslin Hall, Stockport, and in April 1800 the partners attended a meeting at the Castle Inn which proved to be a second great turning point in Radcliffe's career. A year earlier, in 1799, the handloom weavers, who were suffering from one of those depressions which interrupted the prosperity even of the Golden Age, had been demanding as a remedy the restriction of the export of yarn, which was becoming one of the most important branches of British trade.[4] The manufacturing interests, in conjunction with bleachers and printers, had taken up the cause with enthusiasm, and the Stockport meeting had been called to further it. To this appeal Radcliffe responded with religious fervour, but he felt obliged in candour to inform the meeting that he could not find weavers or looms enough to dispose of his own yarn. This seemed a fatal bar to the proposed crusade. But he added that he believed that " with the same hands now employed in weaving we might by the invention of some plan

[1] Radcliffe. *Origin, etc.*, pp. 10–11.
[2] Radcliffe, *op. cit.*, pp. 16, 68.
[3] Radcliffe, pp. 14–15.
[4] Radcliffe, pp. 73–7.

or other . . . ere long be able to consume all the surplus yarns at that time spun for export." He pledged himself to devote his energies to its discovery, on condition that those present would pledge themselves to support the prohibition of the export of yarn if he were successful. This solemn league and covenant was ratified in bumpers of generous wine. It was a great moment—to Radcliffe's mind the greatest in modern history. It was his deepest conviction that " but for the meeting at the Castle Inn and the mutual pledge then given in all probability the new system had never been known." It was not, however, till more than a year later that the two partners, finding themselves with £11,000 to dispose of, " came to the conclusion of purchasing the premises in the Hillgate from Mr. Oldknow and Mr. Arkwright, for the sole purpose of filling them with looms on some new plan, and just so much spinning machinery as would supply the looms with weft." [1]

" Remembering my pledge, and confident that the system was to be found, I shut myself up (as it were) in the mill on the 2nd Jan., 1802, and with joiners, turners, filers, &c., set to work . . . Before the end of the month I began to divide the labour of the weavers, employing one room to dress the whole web, in a small frame for the purpose, ready for the looms in another room, so that the young weaver had nothing to learn but to weave ; and we found this a great improvement for besides the advantage of learning a young weaver in a few days we found . . . that we could make good cloth in the upper rooms with the dressed yarn quite dry which could not be done in the old way of dressing.

" Amongst my old weavers I had a very respectable elderly man residing in Bredbury, who with his family had worked for me many years of the name of Thomas Johnson. He had a son whom I knew to be more ingenious about his loom, than fond of close working . . . I told him . . . we must complete a system similar to the carding and drawing in the spinning line, but the first thing he must attend to was the motion for taking up the cloth as it was wove. For this, as in every step afterwards, he suggested many plans before I would be at the expense of trying any one . . . This motion to the loom being at length accomplished . . . I set Johnson to plan for the warping and dressing, suggesting several ideas myself. His uncommon genius led him to propose many things to me, but I pointed out objections to them all, and set him to work again. His mind was so teased with difficulties that he began to relieve it by drinking for several days together . . . at length we brought out the present plan . . . but the difficulties, expense, and loss in the warping part were immense . . . we got into such a dilemma, that every man I had about me gave up all as lost . . . I listened to all they had to say, as to the difficulties which to them at the time were insurmountable . . . I told them they would all be got over, I then requested them to point out that part of the system which they thought most insurmountable, and that if I got over that difficulty, they would engage to get over the minor ones. To this they all agreed. When in less than an hour, with my own hands, I shewed them that the particular part of the machinery which they wished to throw aside, would not only answer but become an essential part of the system. This raised their spirits and induced them to persevere in the good cause . . . but it was only by dividing the warp, say half the dry beams on one side and half on the other, so as to have

[1] Radcliffe, pp. 11–15.

S

the yarn thin in the brushes that this noiseless simple DRESSING-MACHINE which, when the pledge is redeemed, might annually redeem twenty-five millions sterling, of our national debt . . . became complete.

"We had now been nearly two years at work, all our hands, warpers, dressers, weavers, &c. (3 to 400) had been hired at weekly wages, and owing to the great fall in the price of weaving, I think from 17*s.* to 10*s.* 6*d.* . . . it appeared, on making an estimate, that we had been sinking £70 weekly for many weeks, before we could get our hands upon piece-work. But we had too deeply embarked to think of retreating; and, had I desisted at this trying point, there would not have been a power-loom at work this day in the kingdom.[1]

These experiments had excited a lively interest amongst the cotton manufacturers, and as early as 1802 Radcliffe's factory was visited by two of the most eminent. Sir Robert Peel, who came first, expressed great doubts, but added that if Radcliffe brought out a system by which the surplus yarns might be worked up the Government ought to give him a million of money.[2] Mr. Horrocks, of Preston, gave more positive encouragement by saying, "I see your system will answer, and you only want time to teach your hands to work it." [3] With a view to concealing the new inventions from foreign enquirers, the patents were taken out in 1803–4, by a stroke of Bismarckian policy, in the name of the unknown inventor, who in addition to this unwonted honour received £50 as a consideration,[4] and Radcliffe was not without hopes that the Government would purchase the rights for the use of the trade at large.[5] In this he was disappointed, though he had the satisfaction of seeing his improvements adopted by the leading firms in England and Scotland—by the Peels of Tamworth and of Burton, the Strutts of Derby, by Daintry, Ryle & Co. of Macclesfield, and James Monteith of Glasgow—so that the premiums paid to him amounted to £1,500 in eight months of 1806–7.[6] But this did not compensate him for the absorption of his capital in his factory experiment, which could only succeed on Radcliffe's plan by the maintenance and reconstruction of his large connection with the country weavers. The stoppage of trade caused by the Orders in Council and Napoleon's blockade rendered this impossible, and Radcliffe, who had separated from his partner in 1806, became bankrupt in 1807.[7] His explanation of this and his subsequent troubles as arising from the machinations of the Manchester Foreign Anglo-Junto, who had control over British Ships, Colonies, and Commerce, who instigated the weavers to destroy the power-looms, and who sacrificed the Turkey Company to the Russian Millionists—these and many other curious matters may be found in the "*Narrative containing a candid statement of the strenuous, persevering, and uncompromising endeavours of William Radcliffe to remove the cause which has brought this country to its present crisis, written by himself.*"

This second portion of the title of his book sufficiently indicates the mixture of apostolic zeal and self-complacent prolixity, and of obsession by fixed and limited ideas, which makes the "Narrative" one of the most amusing and least objective of historical documents. Like most such documents, however, it yields ample instruction to those who approach it with a judicious blend of

[1] Radcliffe, pp. 20–3. [2] Radcliffe, p. 24. [3] *Ibid.*, p. 30.
[4] *Ibid.*, p. 26. [5] *Ibid.*, pp. 24, 27. [6] *Ibid.*, p. 28.
[7] *Ibid.*, pp. 112–3.

critical detachment and sympathetic imagination. The clue is provided in the other half of the title, which runs "*Origin of the New System of Manufacture, commonly called 'Power-Loom Weaving,' and the purposes for which this system was invented and brought into use fully explained.*" What was the new system? The unguarded reader naturally assumes that "Power-Loom Weaving" was the essence of it; but that is to miss the whole drift of Radcliffe's argument and belittle his achievement as understood by himself. When he wrote his pamphlet in 1828 the power-loom was coming into general use and he was anxious that the credit of having made it workable should be attributed, as it usually has been, to his Dressing Machine. Nevertheless, he regarded the power-loom as an excrescence on the purity of the New System of Manufacture, which had been invented by himself for the purpose of restoring the Golden Age by re-establishing the cottage weaver, together with his old patron the merchant manufacturer, and "launching our national vessel afresh in the track our forefathers, after many struggles, finally confirmed her in, by the 28th of George the Third."[1]

Radcliffe's New System was not to be a factory system, but a system of manufacture. The use made of the factory in dressing the yarn and training the weavers was to be subordinated to the revival of cottage weaving on a great scale.

In showing his improvements to Mr. Horrocks, of Preston, in 1802, he carefully explained that though he was obliged to bring such a number of boys and girls into the factory to work the looms, yet when the hands had been taught to work them it was his intention to disperse all these looms into the cottages of the weavers throughout the country, in lots proportioned to the number of children in families, delivering them the warps dressed and prepared ready for weaving off the beam and their wefts in cops ready for the shuttle.[2] Horrocks' power-loom, which derived, according to Radcliffe, its only new idea from his own[3] improved handloom, he must have regarded when it came to be generally adopted with much the same feelings as the fisherman felt toward the Afreet whom he had unwittingly released from the bottle. He himself continued to prefer his own handlooms, 200 of which in 1812 "turned off about 600 pieces of printing cambrics per week, never excelled in quality by any looms put in motion by steam"[4] and, though he was obliged in later years to adopt the power-loom in his own business, the last words he addressed to his fellow-countrymen, as an old man of over eighty, were an earnest recommendation that the terrible problems of unemployment and poverty of the year 1840 should be solved by distributing his patent handloom with a dressed or sized warp and weft amongst the cottages of the poor.[5] The unwilling creator of the factory system in weaving was the last of the fully convinced and uncompromising exponents of mercantilism.

Like William Cobbett and many another famous Englishman, William Radcliffe combined revolutionary energy in action with deeply rooted and reactionary conservatism in thought and sentiment.

[1] Radcliffe, Preface to 2nd Edition (1840), pp. vii.–viii.
[2] Radcliffe, p. 30.
[3] *Ibid.*, p. 27.
[4] *Ibid.*, p. 184.
[5] *Ibid.*, Preface, p. viii.

INDEX

INDEX

INDEX

INDEX

INDEX

INDEX

INDEX

INDEX

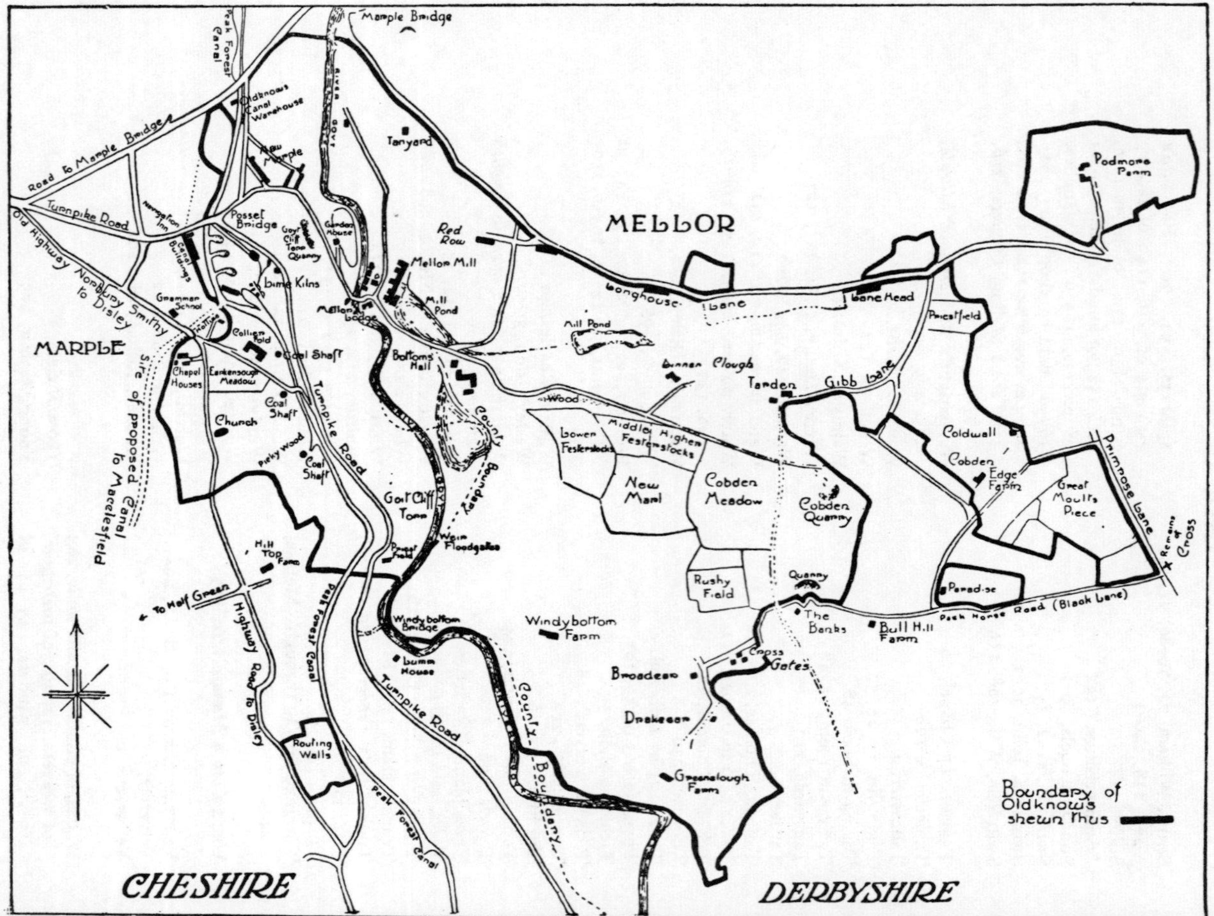
MELLOR
MARPLE
CHESHIRE
DERBYSHIRE
Boundary of Oldknows shewn thus
Marple Bridge
Road to Marple Bridge
Turnpike Road
Old Highway Norbury Smithy to Disley
Peak Forest Canal
Oldknows Canal Warehouse
New Marple
Tanyard
Red Row
Mellor Mill
Mill Pond
Mellor Lodge
Posset Bridge
Canal Buildings
Garden House
Lime Kilns
Grammar School
Chapel Houses
Church
Coal Shaft
Pricky Wood
Bottoms Hall
Goit Cliff Tom
Weir Floodgates
County Boundary
Hill Top Farm
To Hollins Green
Highway Road to Disley
Windybottom Bridge
Windybottom Farm
Turnpike Road
Routing Walls
Longhouse Lane
Lane Head
Priestfield
Mill Pond
Linnen Clough
Tarden
Gibb Lane
Wood
Lower Festerstocks
Middle Higher Festerstocks
New Marl
Cobden Meadow
Cobden Quarry
Coldwall
Cobden Edge Farm
Great Moults Piece
Primrose Lane
Remains of Cross
Rushy Field
Quarry
Paradise
Pack Horse Road (Black Lane)
The Banks
Bull Hill Farm
Cross Gates
Broadeer
Drakeear
Greenslough Farm
Podmore Farm
Site of proposed Canal to Macclesfield